T0418804

Analysis and Synthesis of Polynomial Discrete-Time Systems

An SOS Approach

Analysis and Synthesis of Polynomial Discrete-Time Systems

An SOS Approach

Shakir Saat

Universiti Teknikal Malaysia Melaka, Faculty of Electronic and Computer Engineering, Durian Tunggal, Melaka, Malaysia

Sing Kiong Nguang

The University of Auckland, Department of Electrical and Computer Engineering, Auckland, New Zealand

Alireza Nasiri

Hormozgan University, Department of Electrical and Computer Engineering, Bandar Abbas-Iran

Butterworth-Heinemann is an imprint of Elsevier
The Boulevard, Langford Lane, Kidlington, Oxford OX5 1GB, United Kingdom
50 Hampshire Street, 5th Floor, Cambridge, MA 02139, United States

Library of Congress Cataloging-in-Publication Data
A catalog record for this book is available from the Library of Congress

British Library Cataloguing-in-Publication Data
A catalogue record for this book is available from the British Library

ISBN: 978-0-08-101901-6

For information on all Butterworth-Heinemann publications
visit our website at https://www.elsevier.com/books-and-journals

Publisher: Mara Conner
Acquisition Editor: Sonnini R. Yura
Editorial Project Manager: Ana Claudia A. Garcia
Production Project Manager: Sruthi Satheesh
Designer: Victoria Pearson

Typeset by VTeX

Contents

About the Authors

Shakir Saat is an Associate Professor and Deputy Dean (Academic) in the Faculty of Electronic and Computer Engineering at Universiti Teknikal Malaysia Melaka, Malaysia. He obtained his B.E and M.E in Electrical Engineering from Universiti Teknologi Malaysia, Malaysia, and the PhD from the University of Auckland, New Zealand. His research interests are nonlinear system control theory and wireless power transfer technologies.

Sing Kiong Nguang is a chair professor in the Department of Electrical and Computer Engineering at the University of Auckland, Auckland, New Zealand. He received the B.E. (with first class honors) and the PhD degree from the Department of Electrical and Computer Engineering of the University of Newcastle, Callaghan, Australia. He has published over 300 refereed journal and conference papers on nonlinear control design, nonlinear control systems, nonlinear time-delay systems, nonlinear sampled-data systems, networked control systems, biomedical systems modeling, fuzzy modeling and control, biological systems modeling and control, and food and bioproduct processing. He has/had served on the editorial board of a number of international journals. He is the Chief-Editor of the International Journal of Sensors, Wireless Communications and Control.

Alireza Nasiri received the BSc degree in Electrical Engineering from the University of Tehran, Tehran, Iran, the MSc degree in Electrical Engineering from the Iran University of Science and Technology, Tehran, and the PhD degree from the University of Auckland, Auckland, New Zealand. He is currently an Assistant Professor with the Department of Electrical and Computer Engineering, Hormozgan University, Bandar Abbas, Iran, and the Vice President of Hormozgan Science and Technology Park, Hormozgan, Iran. His current research interests include nonlinear control, active noise control, and hyphenate control.

Analysis and Synthesis of Polynomial Discrete-Time Systems. DOI: 10.1016/B978-0-08-101901-6.00014-1

Preface

The polynomial discrete-time systems are the type of systems where the dynamics of the systems are described in polynomial forms. This system is classified as an important class of nonlinear systems due to the fact that many nonlinear systems can be modeled as, transformed into, or approximated by polynomial systems.

The focus of this book is to address the problem of controller design for polynomial discrete-time systems. The main reason for focusing on this area is because the controller design for such polynomial discrete-time systems is categorized as a difficult problem. This is due to the fact that the relation between the Lyapunov matrix and the controller matrix is not jointly convex when the parameter-dependent or state-dependent Lyapunov function is under consideration. Therefore the problem cannot possibly be solved via semidefinite programming (SDP). In light of the aforementioned problem, we establish novel methodologies of designing controllers for stabilizing the systems both with and without H_∞ performance and for the systems with and without uncertainty. In this book, we consider two types of uncertainty, polytopic uncertainties and norm-bounded uncertainties. A novel methodology for designing a filter for the polynomial discrete-time systems is also developed. We show that through our proposed methodologies, a less conservative design procedure can be rendered for the controller synthesis and filter design.

In particular, we propose a so-called integrator method, where an integrator is incorporated into the controller and filter structures. In doing so, the original systems can be transformed into augmented systems. Furthermore, the state-dependent Lyapunov function is selected in a way that its matrix is dependent only upon the original system state. Through this selection, a convex solution to the controller design and the filter design can be efficiently obtained. However, the price we pay for incorporating the integrator into the controller and filter structures is a large computational cost, which prevents us from using this method in general. To reduce the computational requirements for our design methodologies, we consider a number of simpler classes of polynomial systems.

Based on this integrator approach, we first consider the state feedback control problem. In this case, the nonlinear state feedback control is first tackled and followed by the robust control problem in which the uncertain terms are described as polytopic forms. The robust control problem with norm-bounded uncertainty is next studied. Then, we discuss the nonlinear H_∞ state feedback control problem and robust nonlinear H_∞ state-feedback control problem with polytopic and norm-bounded uncertainty. The design ensures that the ratio of the regulated output energy and the disturbance energy is less than a prescribed performance level. The filter design is next tackled and followed by the output feedback control problem. In the output feedback control, the problem of system uncertainties and disturbances are addressed. The existence of such controllers and a filter are given in terms of the solvability of polynomial matrix inequalities (PMIs). The problem is then formulated as sum-of-squares (SOS) constraints; therefore, it can be solved by any SOS solver. In this book, SOSTOOLS is used as an SOS solver.

Motivated by most of the existing control design methods for discrete-time fuzzy polynomial systems cannot guarantee their Lyapunov function to be a radially unbounded polynomial function, and hence the global stability cannot be ensured. This book also provides controller design methods for discrete-time fuzzy polynomial systems that guarantee a radially unbounded polynomial Lyapunov function that ensures the global stability.

Finally, to demonstrate the effectiveness and advantages of the proposed design methodologies in this book, numerical examples are given in each designed control system. The simulation results show that the proposed design methodologies can stabilize the systems and achieve the prescribed performance requirements.

Shakir Saat
Universiti Teknikal Malaysia Melaka, Faculty of Electronic and Computer Engineering, Durian Tunggal, Melaka, Malaysia

Sing Kiong Nguang
The University of Auckland, Department of Electrical and Computer Engineering, Auckland, New Zealand

Alireza Nasiri
Hormozgan University, Department of Electrical and Computer Engineering, Bandar Abbas, Iran

15 March 2017

Chapter 1

Introduction

CHAPTER OUTLINE

1.1 NONLINEAR SYSTEMS

Nonlinear systems play a vital role in the control systems from an engineering point of view. This is due to the fact that in practice all plants are nonlinear in nature. This is the main reason for considering the nonlinear systems in our work. In mathematics, a nonlinear system does not satisfy the superposition principle, or its output is not directly proportional to its input. The best example to explain nonlinearity is obviously a saturation. This condition exists because it is impossible to deliver an infinite amount of energy to any real-world system.

In general, the state and output equations for nonlinear systems may be written as follows:

$$\begin{aligned} \dot{x}(t) &= f[x(t), u(t)], \\ y(t) &= g[x(t), u(t)]. \end{aligned} \tag{1.1}$$

Analysis and Synthesis of Polynomial Discrete-Time Systems. DOI: 10.1016/B978-0-08-101901-6.00001-3

The Lorenz chaotic system is an example of a nonlinear system described as follows:

$$\begin{aligned}\dot{x}_1(t) &= -10x_1(t) + 10x_2(t) + u(t),\\ \dot{x}_2(t) &= 28x_1(t) - x_2(t) + x_1(t)x_3(t),\\ \dot{x}_3(t) &= x_1(t)x_2(t) - \frac{8}{3}x_3(t).\end{aligned} \tag{1.2}$$

Notice that because of the terms $x_1(t)x_3(t)$ and $x_1(t)x_2(t)$, system (1.2) is nonlinear in nature. In the sequel, the nonlinear discrete-time systems considered in this book are introduced.

1.2 NONLINEAR DISCRETE-TIME SYSTEMS

Nowadays we can see that almost all controllers are implemented using computers. Such controllers are known as digital controllers. Basically, the use of digital controllers has rapidly increased since the first idea of using digital computers as one of the components in control systems emerged somewhere in 1950. The detailed history of this development can be found in [1]. The main reason for this development is due to the advances in hardware; hence it provides the control engineer with more powerful, reliable, faster, and above all cheaper computers that could be implemented as process controllers. The another significant factor that drives the increase in development of digital controllers is the advantage of working with digital signals rather than continuous-time signals [2]. The aforementioned factors generally motivate us to deliver the research in the framework of discrete-time systems rather than continuous-time systems.

Generally, a closed-loop system of computer controlled systems can be illustrated by Fig. 1.1, where the output of the process $y(t)$ is a continuous-time signal. The measurements of the output signal are fed into an analog-to-digital (A-D) converter, where the continuous-time signal is converted into a digital signal, a sequence of measurements at sampling times t_k. At this point, if a digital measurement device is used, the A-D converter is no longer needed. This is true because the measurements are now taken at sampling times only. The computer interprets the converted output signals $y(t_k)$ as a sequence of numbers, and this sequence is then used by the control algorithm to compute a sequence of digital control signals $u(t_k)$. Notice that the process input is in continuous-time, and hence a digital-to-analog (D-A) converter is used to transform the signals into a continuous-time signal. It is important to highlight here that between the sampling instants the system is in open-loop mode. The system is synchronized by a real time clock in the computer. Consequently, the inter-sample behavior is very often an issue

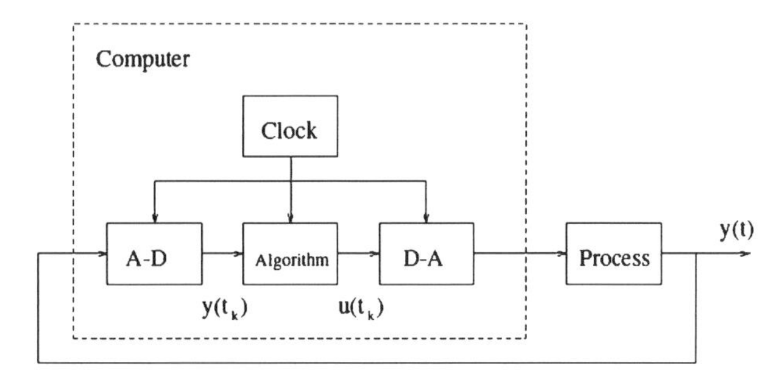

FIGURE 1.1 Schematic diagram of a computer-controlled system.

and should not be disregarded. However, in many applications it is sufficient to describe the dynamic behavior of the system at the sampling instants. At this stage, the interested signals are only at discrete time, and this system is classified as a discrete-time system [1,3]. We can now simply justify that if the dynamic of the process is in linear forms, then such a system is called a linear discrete-time system. Meanwhile, if the behavior of the process is nonlinear, then it is known as a nonlinear discrete-time system.

1.2.1 Discretization

Since systems in this world are naturally in continuous time, discretization shall be performed so that an approximated discrete-time system can be obtained. We further list the available methods in the discretization framework:

- **Euler's forward differentiation method and Euler's backward differentiation method**: The methods are based on approximations of the time derivatives of the differential equation. The forward method is commonly used in developing simple simulators, whereas the backward method is normally used in discretizing simple signal filters and industrial controllers. The forward differentiation method is somewhat less accurate than the backward one, but it is simpler to use. Particularly, with nonlinear models, the backward differentiation may give problems since it results in an impact equation for the output variable. In contrast, the forward differentiation method always gives an explicit equation to the solution.
- **Zero-Order Hold (ZOH) method:** Using this method, it is assumed that the system has a zero-order hold element on the input element of the system. This is the case where the physical system is controlled by a computer via a digital-to-analogue (D-A) converter. ZOH means that the physical input signal to the system is held fixed between the discrete points. Unfortunately, this method is relatively complicated to apply, and

in practice the computer tool MATLAB or LabVIEW can perform the job.

- **Tustin's method:** The discretization method is based on an integral approximation where the integral is interpreted as the area between the integrand and the time axis, and this area is approximated with trapezoids. It should be noted here that in Euler's method this area is approximated by a rectangle.
- **Tustin's method with frequency prewarping, or Bilinear transformation:** This method follows Tustin's method with a modification so that the original continuous-time system and the resulting discrete-time system have exactly the same frequency response at one or more specified frequencies.

It should be mentioned here that discretization methods are not the main focus of this book. The discretization is only applied in the simulation examples as to convert the continuous-time systems into discrete-time systems. To perform such a discretization, in this research work, the Euler method is used due to its simplicity.

1.2.2 Brief overview on the literature of nonlinear discrete-time systems

Due to the tremendous increase in digital control applications, a theory for discrete-time systems must be one of the important theories to be investigated especially for control design purposes. It is obvious that the desired performance may not be achieved if the controller design is based on the linearized model, and in many cases, it is not possible to control nonlinear systems from the linearized model. Besides, the linear control theory cannot be applied in the cases where a large dynamic range of process variables is possible, multiple operating points are required, the process is operating close to its limits, small actuators cause saturation, etc. [3]. The feedback linearization approach [4] also cannot be extended to handle a system with parametric uncertainties. This is the major drawback of the feedback linearization approach. With this knowledge, a significant amount of works can be found in the literature that attempts to provide a more general and less conservative result than the linearized approach. One of the popular approaches is obviously the backstepping control technique [5]. This approach is actually a combination of two popular theories, Lyapunov stability theory and the geometric method. By exploring the recursive design procedure the time-varying uncertainties and parameter uncertainties can be incorporated into the problem formulation. However, it is difficult to find a general class of Lyapunov functions that could ensure the stability of such systems. This is the main disadvantage of backstepping control techniques, and obviously this drawback is common to all approaches that use constructive procedures in developing Lyapunov candidates [6].

Besides the existence of feedback linearization and backstepping control techniques for stabilizing nonlinear systems, there is one more popular method available that is widely used in control system engineering and is called gain-scheduling [7]. The primary advantage of gain-scheduling for nonlinear control design is that it is usually possible to meet performance objectives over a wide range of operating conditions while still taking advantage of the wealth of tools and designers' experience from linear controller synthesis. From this gain-scheduling approach a more systematic control design technique is developed in the framework of linear parameter-varying (LPV) systems that guarantees stability and performance properties [8–10]. However it is important to highlight here that the stability and performance properties of the LPV systems only hold locally, and it is well known that the application of LPV control techniques always requires convertion of the nonlinear systems into their quasi-LPV forms. These are usually the main sources of conservatism of this gain-scheduling method. Another popular approach in this area is based on the Takagi–Sugeno fuzzy approach. It is well known that Takagi–Sugeno fuzzy models can be used to approximate nonlinear systems [11–25]. However, in the T–S fuzzy model, the premise variables are assumed to be bounded. In general, the premise variables are related to the state variables, which implies that the state variables have to be bounded. This is one of major drawbacks of the T–S fuzzy model approach.

Based on the above statements, it is clear that there is plenty of room available to conduct study on stabilizing nonlinear discrete-time systems, and a better methodology should be proposed to reduce the conservatism of the above-mentioned approaches. This motivates us to deliver the research in the framework of discrete-time systems, so that a less conservative approach can be proposed for stabilizing nonlinear discrete-time systems. Before ending this section, it is important to note that general nonlinear discrete-time systems are too complex, and hence in this research work we limit the scope by considering only the polynomial discrete-time systems. The reason of selecting the polynomial systems will be given in the following text.

1.3 POLYNOMIAL SYSTEMS

It is well known that a wide class of nonlinear systems can be exactly represented by polynomial systems, that is, Lorenz chaotic systems. Moreover, the polynomial system has an ability to approach any analytical of nonlinear systems. These advantages explain why the polynomial systems constitute an important class of nonlinear systems and have attracted considerable attention from control researchers to involve themselves in this area, especially on the stability analysis and controller synthesis of polynomial systems [26].

The polynomial systems are systems where the dynamic of the system is given in terms of polynomial functions or polynomial matrices. The general polynomial systems can be described as follows:

$$\begin{aligned} \dot{x}(t) &= f(x(t), u(t)), \\ y(t) &= g(x(t)), \end{aligned} \tag{1.3}$$

where $f(x(t), u(t))$ and $g(x(t))$ are in polynomial forms, and $x(t)$, $u(t)$, and $y(t)$ are respectively the states, input, and the measured output.

Meanwhile, in discrete-time, (1.3) can be written as follows:

$$\begin{aligned} x(k+1) &= f(x(k), u(k)), \\ y(k) &= g(x(k)), \end{aligned} \tag{1.4}$$

where $f(x(k), u(k))$ and $g(x(k))$ are in polynomial forms, and $x(k)$, $u(k)$, and $y(k)$ are respectively the states, input, and the measured output of the system at sampling time k. More precisely, the class of polynomial systems under consideration in this research work is described in terms of a state-dependent linear form as follows:

$$\begin{aligned} \dot{x}(t) &= A(x(t))x(t) + B(x(t))u(t), \\ y(t) &= C(x(t))x(t), \end{aligned} \tag{1.5}$$

where $x(t) \in R^n$ is the state vector, $u(t) \in R^m$ is the input, and $y(t)$ is the measured output. $A(x(t))$, $B(x(t))$, and $C(x(t))$ are polynomial matrices of appropriate dimensions. Notice that (1.5) looks similar to the general nonlinear system, except the matrices $A(x(t))$, $B(x(t))$, and $C(x(t))$ must be of polynomial forms.

In our work, the polynomial discrete-time system is described as follows:

$$\begin{aligned} x(k+1) &= A(x(k))x(k) + B(x(k))u(k), \\ y(k) &= C(x(k))x(k), \end{aligned} \tag{1.6}$$

where $x(k) \in R^n$ is the state vector, $u(k) \in R^m$ is the input, $y(k)$ is the measured output, and $A(x(k))$, $B(x(k))$, and $C(x(k))$ are polynomial matrices of appropriate dimensions.

We provide two examples of polynomial systems.

Example 1. A Lorenz Chaotic System

$$\begin{aligned}\dot{x}_1(t) &= -10x_1(t) + 10x_2(t) + u(t),\\ \dot{x}_2(t) &= 28x_1(t) - x_2(t) + x_1(t)x_3(t),\\ \dot{x}_3(t) &= x_1(t)x_2(t) - \frac{8}{3}x_3(t).\end{aligned} \tag{1.7}$$

Example 2. A Tunnel Diode Circuit

$$\begin{aligned}C\dot{x}_1(t) &= -0.002x_1(t) - 0.01x_1^3(t) + x_2(t),\\ L\dot{x}_2(t) &= -x_1 - Rx_2(t) + u(t),\end{aligned} \tag{1.8}$$

where C is a capacitor value, L is inductance, and R is a resistor.

We can see here that systems (1.7) and (1.8) are in fact nonlinear systems. These two examples illustrate the validity of the statement we claimed earlier that many nonlinear systems can be represented by polynomial forms. It is also important to stress here that in this research we do not focus on the method of discretizing the nonlinear systems to yield their discrete-time version, but our main focus is to perform the controller synthesis for polynomial discrete-time systems.

1.3.1 Recent work on polynomial systems

Our focus in this research work is on the controller synthesis of polynomial discrete-time systems. However, in this section, the results of polynomial continuous-time systems are also discussed. This is due to the fact that some results of the discrete-time systems are based or extended from the continuous-time systems. In this regard, we present the following recent development on the controller synthesis for polynomial systems. It is worth mentioning that we limit the results to the sum-of-squares (SOS) decomposition method and linear matrix inequalities (LMIs) only. This is because the SOS decomposition method will be considered in this research work, and it is in fact complementary to the LMI approach. A detailed description regarding the SOS decomposition method will be provided later in this chapter.

The controller synthesis or stabilization problem is an important area in the research of polynomial systems. Therefore, considerable attention has been devoted to this framework; for instance, see [28–33,36–39]. In the present work, numerous techniques have been proposed to address the controller design problem for polynomial systems. We now briefly describe the proposed techniques.

1. **Dissipation inequalities and SOS**. Dissipativity theory is known to be one of the most successful methods of analyzing and synthesizing the nonlinear control systems [27]. Mathematically speaking, this method is known as dissipation inequalities and has major advantages on the analysis and design of nonlinear systems. This might be due to the fact that the investigation of a possibly large number of differential equations given by the control system description is reduced to a small number of algebraic inequalities. Hence, the complexity of analysis and design is usually essentially reduced. In [28] the dissipation inequalities, together with the SOS programming, have been utilized to stabilize such polynomial systems. In particular, the authors represent their systems to be of descriptor systems or differential-algebraic systems where the functions are described by polynomial functions. They have managed to obtain affine dissipation inequalities by the proposed method, and hence the inequalities can be solved computationally via SOS programming. However, the process of achieving affine dissipation inequalities varies for different types of problem. This means that the proposed method might not work for other problems.
2. **Kronecker products and LMIs**. The stabilization of polynomial systems using Kronecker products method can be found in [29–31], where the polynomial systems have been simplified using the Kronecker product and power of vectors and matrices. Moreover, a new stability criterion for polynomial systems has been developed. A sufficient condition for the existence of the proposed controller is given in terms of LMIs. The proposed controller can be applied to high-order polynomial systems. This is the main advantage of this method. The strength of this approach comes from solid theoretical results on the Kronecker products and the power of vectors and matrices.
3. **Semitensor products**. The semitensor product of matrices is a generalization of the conventional matrix product in the case where the column number of the first factor matrix is not the same as the row number of the second factor matrix. A brief survey for the related material can be found in [32]. The advantage of this method is that general polynomial systems can be considered without any homogeneous assumption. In [32], a method to stabilize the polynomial systems has been developed. In the present paper, we first propose a sufficient condition for a polynomial to be positive definite. Then, the formula for the time derivation of a candidate polynomial Lyapunov function with respect to a polynomial systems is provided. Through the sufficient condition, the candidate of the Lyapunov function can be checked for positive definiteness and for negative definiteness of its derivative. A sufficient condition is given by a system of linear algebraic inequalities. However, using this method,

it is difficult to choose a suitable candidate for the Lyapunov function because there is no unique way to choose that Lyapunov function. An incorrect selection of the Lyapunov candidate leads linear algebraic inequalities that have no solution. This is the main challenge of applying this method in the framework of controller synthesis for polynomial systems.

4. **Theory of moments and SOS**. Interesting work on the polynomial stabilization that utilizes the theory of moments can be found in [33]. It has been known for a long time that the theory of moments is strongly related to and is in fact in duality with the theory of nonnegative polynomials and Hilbert's 17th problem on the representation of nonnegative polynomials [34]. In the light of this duality relationship, the authors of [33] study the problem of polynomial system stabilization. They managed to show that the global solution to the problem can be obtained in a less conservative way than the available approaches, and the solution can be solved easily by SDP [35]. However, to achieve a convex solution to the controller synthesis problem, the Hermite stability criterion is used rather than the Lyapunov stability theorem. In doing so, the controller matrix can be decoupled from the Lyapunov matrix, and the solvability conditions of the proposed controller are developed through a hierarchy of convex LMI relaxations. As stated by the authors, this methodology suffers from a large number of constraints in a PMI, which consequently leads to the need for reliable numerical software to handle the problem.
5. **Fuzzy method and SOS**. The T–S fuzzy model is well known to be good at approximating such nonlinear systems. Using this approach, [36] presented an SOS approach for modeling and control of nonlinear dynamical systems using polynomial fuzzy systems. A polynomial Lyapunov function has been proposed in this work rather than a quadratic Lyapunov function. Hence the result is more general and less conservative than available LMI-based approaches of T–S fuzzy modeling and control. Furthermore, a sufficient condition of the existence of a controller is given by polynomial matrix inequalities and formulated as SOS constraints. On the other hand, in [37] an improved sum-of-squares (SOS)-based stability analysis result is proposed for the polynomial fuzzy-model-based control system, formed by a polynomial fuzzy model and a polynomial fuzzy controller connected in a closed loop. Two cases, perfect and imperfect premise matching, are considered. Under the perfect premise matching, the polynomial fuzzy model and polynomial fuzzy controller share the same premise membership functions. When different sets of membership functions are employed, it falls into the case of imperfect premise matching. Based on the Lyapunov stability theory, improved SOS-based stability conditions are derived to

determine the system stability and facilitate the controller synthesis approach. The application of rge polynomial T–S fuzzy approach to the two-link robot arm can be found in [38]. Meanwhile, for static output control, a result can be found in [39]. However, in the T–S fuzzy model, the premise variables are assumed to be bounded. In general, the premise variables are related to the state variables, which implies that the state variables have to be bounded. This is one of the major drawbacks of the T–S fuzzy model approach.

6. **Lyapunov method and SOS**. This is a common method widely applied in the literature for stabilizing polynomial systems. This method is used in this research work, and therefore the complete literature of this framework is provided further.

1.3.1.1 On the literature on controller synthesis for polynomial systems: the Lyapunov method and SOS decomposition approach

It is well known that Lyapunov's stability theory [40] is one of the most fundamental pillars in control theory. Although this method was introduced more than hundred years ago, it remains popular among control researchers. This success is owed to its simplicity, generality, and usefulness. The Lyapunov stability is a method that was developed for analysis purposes. However, it has become of equal importance for control designs over the last decades [6–48]. The Lyapunov stability theory can be generalized as follows. Let us consider the problem of solving the stability for an equilibrium of a dynamical system $\dot{x} = f(x)$ using the Lyapunov function method. It is clear that to find a stability using the Lyapunov method, we need to find a positive definite Lyapunov function $V(x)$ defined in some region of the state space containing the equilibrium point whose derivative $\dot{V} = \frac{dv}{dx} f(x)$ is negative semidefinite along the system trajectories. Taking the linear case, for instance, $\dot{x} = Ax$, these conditions amount to finding a positive definite matrix P such that $A^T P + PA$ is negative definite [49]; then the associated Lyapunov function is given by $V(x) = x^T Px$. Meanwhile, for discrete-time systems $x(k+1) = f(x(k))$, we need to search for a positive definite Lyapunov function $V(x)$ defined in some region of the state space containing the equilibrium point whose difference of the Lyapunov function, $\Delta V = x^T(k+1)Px(k+1) - x^T(k)Px(k)$, is negative semidefinite along the system trajectories. The associated Lyapunov function is given by $V(x) = x^T Px$.

Since the SOS decomposition technique introduced about 10 years ago [50], the system analysis for polynomial systems can be performed more efficiently because it helps to answer many difficult questions on system analysis that were hard to answer before. The popularity of this method grew quickly among the community of control researchers because the algorithmic analysis of nonlinear systems can be delivered using the most popular Lyapunov method (as discussed earlier). Generally, the most interesting and important point that was never seen until recently is that the amount of proving the certificates of the Lyapunov function $V(x)$ and $-\dot{V}(x)$ can be reduced to the SOS [50]. Notice that, for small systems, the construction of the Lyapunov function can be done manually. The difficulty of this construction is solely dependent upon the analytical skills of the researcher. However, when the vector field of the system $f(x)$ and the Lyapunov function candidate $V(x)$ are in polynomial forms, then the Lyapunov conditions are essentially polynomial nonnegativity conditions, which can be NP-hard to test [65]. This is probably due to the lack of algorithmic constructions of Lyapunov functions. However, if these nonnegativity conditions are replaced by the SOS conditions, then not only testing the Lyapunov function conditions, but also constructing the Lyapunov function can be done effectively using SDP [50]. This is the main advantage of using the SOS decomposition approach because the solution is indeed tractable. We will further describe the details of the SOS decomposition method.

The recent results in the framework of state feedback control synthesis for polynomial systems which utilize SOS decomposition method can be referred in [51–54]. In particular, [51,52] propose the polynomial systems to be represented as a state-dependent linear form, and a state-dependent Lyapunov function is proposed to be in terms of polynomial vector fields. The introduction of a state-dependent Lyapunov function or parameter-dependent Lyapunov function arises due to the fact that a quadratic Lyapunov function is always inadequate to stabilize the polynomial systems. Furthermore, sufficient conditions to the problem are formulated as state-dependent LMIs and solved using the SOS-SDP-based programming method. It is well known that optimizing the control problem for polynomial systems is hard because the solution is always not jointly convex. In the present paper, such a nonconvexity is avoided by assuming the Lyapunov matrix $P(x)$ to be dependent upon the states $x(t)$ whose dynamics are not directly affected by the control input, i.e., the states whose corresponding rows in the input matrix $B(x)$ are zero. This, however, leads the result to be conservative. More recent and less conservative results can be found in [53]. In this paper, the effect of the nonlinear terms that exist in the problem formulation is described as an index, so that the control problem can be transformed into a tractable solution and can be possibly solved via SDP [35]. The optimization approach is proposed to find a zero optimum of this index and solved using SOS programming effectively. However, to render a convex solution, the authors follow the same assumption as made in [52]. An improved version of the aforementioned approach can be found in [54], where an additional matrix variable is introduced to decouple the Lyapunov matrix from the system matrices. Therefore, the controller design can be performed in a more relaxed way, and the proposed methodology can be extended to the robust control problem of polynomial systems. However, to obtain a convex solution, the nonconvex term is bounded from above. Therefore, the stability can only be guaranteed within the bound region.

Sometimes it is difficult to synthesize a controller that works globally. Besides, in a restricted region, local controllers often provide a better solution than global controllers. Some developments in this field can be found in [55, 56]. In [55], a rational Lyapunov function of states was used to synthesize the polynomial systems. The variation of states is bounded, and the domain of attraction was embedded in the specified region by the nonlinear vector. With this, the state feedback controller is established and formulated as a set of polynomial matrix inequalities and solved using any SOS programming. The coupling between system matrices and the Lyapunov matrix causes the results to be quite conservative in general. Hence, [56] relaxes this issue by introducing a slack variable matrix. In doing so, the Lyapunov matrix is decoupled from the system matrices. Now, the parameterization of the resulting controllers is independent of the Lyapunov matrix variables. This allows them to extend their result to construct robust controller for uncertain polynomial systems using state-dependent Lyapunov functions.

With the knowledge that the full-state variables are not always accessible in practical nonlinear systems and the dynamic output feedbacks result in high-order controllers, which may not be practical in industry, the static output feedback design attracts much attention among practitioners. Some developments of this area that utilized the SOS decomposition approach can be found in [57–60]. The systems discussed in [57] are represented in a state-dependent linear form. More precisely, the authors assumed that the control input matrix has some zero rows and the Lyapunov function only depends on states whose corresponding rows in control matrix are zeros, that is, the state dynamics are not directly affected by the control input. This assumption leads to the conservatism of the controller design. The latest results of this area can be found in [58–60], where an iterative algorithm based on SOS has been proposed to convert the nonconvex problem into a convex problem of polynomial system synthesis, so that it can be efficiently solved using SDP. The authors in [58–60] have managed to show that their approach is less conservative than the available approaches and provide more general results in this field. But the main disadvantage of this approach is the selection of the initial polynomial function $\epsilon(x)$, which is hard to choose because it is unknown.

The above-mentioned results are dedicated to solving the polynomial continuous-time systems. In regard to the polynomial discrete-time systems, there are only few results available, which utilize the SOS decomposition method in their approach. The first result is proposed in [61], where the authors employ a state-dependent polynomial Lyapunov function as their Lyapunov candidate. Then, some transformations are required to represent the system with introduction of new matrices (in polynomial). Furthermore, YALMIP and PENOPT for PENBMI [62,63] have been utilized to solve the problem. However, the main drawback of this approach is that the selected new matrices are not unique and hence difficult to choose. The most recent result was addressed by [64], where the nonconvex term is bounded from above; the optimization is carried out to find a zero optimum for the nonlinear term. Here, the problem was formulated as SOS and could be solved by using any SOS solver. By bounding the nonconvex terms the controller that resulted from this method can only guarantee the closed-loop stability within bounds. This is similar to the method proposed in [53,54]. Therefore, they share the same weaknesses as encountered in [53,54].

1.3.2 Sum-of-squares (SOS) decomposition

In this section, we give a brief overview of the SOS decomposition method. A more detailed description of the SOS decomposition method can be referred in [50].

Generally, proving the nonnegativity of multivariable functions is considered as one of the important aspects in control system engineering. This problem is similar to the problem of proving the nonnegativity of a Lyapunov function. If the nonlinear system is concerned, then it is hard to prove the nonnegativity of such systems. Basically, the problem is to prove that

$$F(x_1, \ldots, x_n) \geq 0, \quad x_1, \ldots, x_n \in \Re. \tag{1.9}$$

A great amount of research has been devoted to proving (1.9). However, up to now, there is no unique solution to the problem in (1.9).

Thus, some limit should be applied to the possible functions $F(\cdot)$, at the same time making the problem general enough to guarantee the applicability of the results. It has been shown in [50] that considering the case of polynomial functions is a good compromise for this issue.

Definition 1. [50] A form is a polynomial if all the monomials have the same degree $d := \sum_i \alpha_i$. In this case, the polynomial is homogeneous of degree d since it satisfies $f(\lambda x_1, \ldots, \lambda x_n) = \lambda^d f(x_1, \ldots, x_n)$.

It should be highlighted here that the general problem of testing global positivity of a polynomial function is NP-hard problem (when the degree is at least four) [65]. Therefore, a problem with a large number of variables will have unacceptable behavior for any method that guarantees obtaining the right answer in every possible instance. This is actually the main drawback of theoretically powerful methodologies such as the quantifier elimination approach [66,67].

The question now is: are there any conditions to guarantee the global positivity of a tested polynomial time function? This question underlines the existence of the SOS decomposition approach [50] as one condition to guarantee the global positivity of polynomial functions.

It is obvious that a necessary condition for a polynomial $F(x)$ to satisfy (1.9) is that the degree of the polynomial in the homogeneous case must be even. Hence, a simple sufficient condition for a real-valued function $F(x)$ to be positive everywhere is given by the existence of an SOS decomposition

$$F(x) = \sum_i {f_i}^2(x). \tag{1.10}$$

It can been seen that if $F(x)$ can be written as (1.10), then the nonnegativity of $F(x)$ can be guaranteed. It is stated in [50] that for the problem to make sense, some restriction on the class of functions f_i has to be again imposed. Otherwise, we need to always define f_1 to be the square root of F, but this results in a condition both useless and trivial.

It has been shown in [50] that $F(x)$ is an SOS polynomial if and only if there exists a positive definite matrix Q such that

$$F(x) = z^T Qz, \tag{1.11}$$

where $z(x)$ is the vector of all monomials of degree less than or equal to the half degree of $F(x)$. This is the idea given in [71], and it can be shown to be conservative in general. The main reason is that since the variables z_i are not independent, representation (1.11) may not be unique, and Q may be positive definite or positive semidefinite for some representations but not for others. Similar issues appear in the analysis of quasi-LPV systems; we refer to [72]. However, using identically satisfied constraints that relate the z_i variables among themselves, it is easily shown that there is a linear subspace of matrices Q that satisfy (1.11). If the intersection of this subspace with the positive semidefinite matrix cone is nonempty for the original function, then F is guaranteed to be SOS and therefore positive semidefinite. So, if F can indeed be written as the SOS of polynomials, then expanding in monomials will provide representation (1.11). The following example explains this concept.

Example 1. [50] Consider the quartic form in two variables

$$
\begin{aligned}
F(x_1, x_2) &= 2x_1{}^4 + 2x_1{}^3x_2 - x_1{}^2x_2{}^2 + 5x_2{}^4 \\
&= \begin{bmatrix} x_1{}^2 \\ x_2{}^2 \\ x_1x_2 \end{bmatrix}^T \begin{bmatrix} 2 & 0 & 1 \\ 0 & 5 & 0 \\ 1 & 0 & -1 \end{bmatrix} \begin{bmatrix} x_1{}^2 \\ x_2{}^2 \\ x_1x_2 \end{bmatrix} \\
&= \begin{bmatrix} x_1{}^2 \\ x_2{}^2 \\ x_1x_2 \end{bmatrix}^T \begin{bmatrix} 2 & -\lambda & 1 \\ -\lambda & 5 & 0 \\ 1 & 0 & -1+2\lambda \end{bmatrix} \begin{bmatrix} x_1{}^2 \\ x_2{}^2 \\ x_1x_2 \end{bmatrix}
\end{aligned} \tag{1.12}
$$

and define $z_1 := x_1{}^2$. $z_2 := x_2{}^2$, $z_3 := x_1x_2$. Take, for instance, $\lambda = 3$. In this case,

$$
Q = L^T L, \; L = \frac{1}{\sqrt{2}} \begin{bmatrix} 2 & -3 & 1 \\ 0 & 1 & 3 \end{bmatrix}. \tag{1.13}
$$

Therefore we have the sum-of-squares decomposition

$$
F(x_1, x_2) = \frac{1}{2}((2x_1{}^2 - 3x_2{}^2 + x_1x_2)^2 + (x_2{}^2 + 3x_1x_2)^2) \tag{1.14}
$$

Parrilo [50] also observed that the existence of (1.11) can be cast as a semidefinite programming [35]. This is the most important property that distinguishes its from other approaches. This feature is proved to be critical in the application to many control-related problems. How does it works? Basically, by expanding the $z^T Qz$ and equating the coefficient of the resulting monomials to the ones in $F(x)$ we obtain a set of affine relations in the elements of Q. We know that, for $F(x)$ being an SOS is equivalent to $Q \geq 0$, then the problem of finding Q which proves that $F(x)$ is an SOS can certainly be cast as a semidefinite program.

Thus, although checking the nonnegativity of $F(x)$ is NP-hard when the degree in $F(x)$ is 4 as stated before, checking whether $F(x)$ can be written as SOS is definitely tractable; it can be formulated as a semidefinite program, which has worst-case polynomial time complexity as mentioned in the previous paragraph. Authors in [50] produced significance results in suggesting that the relaxation is not too conservative in general. It must be noted here that as the degree of $F(x)$ increases or the number of its variables increases, the computational complexity for testing whether $F(x)$ is an SOS significantly increases. Nonetheless, the complexity overload is still a polynomial function of these parameters.

In general, the conversion from SOS decomposition to the semidefinite programming can be manually done for small size instance or tailored for specific problem classes. However, such a conversion is cumbersome in general. Thus the software is absolutely necessary to aid in converting them. Specifically, the relaxation uses Gram matrix methods to efficiently transform the NP-hard problem into LMIs [49]. These can in turn be solved in polynomial time with semidefinite programming (SDP) [49,35]. To date, there exist several freely available toolboxes to formulate these problems in Matlab, for example, SOSTOOLS [73], YALMIP [74], CVX [75], and GLoptiPoly [76]. Whereas SOSTOOLS is specifically designed to address polynomial nonnegativity problems, the latter toolboxes have further functionality, such as modules to solve the dual of the SOS problem, the moment problem.

In this work, we use SOSTOOLS to perform this conversion for our problem formulation. Hence we will describe the working principles of this software in the following section.

Basically, the polynomial case is a well-analyzed problem, first studied by David Hilbert more than century ago [68]. He raised a very popular and important question in his famous list of twenty-three unsolved problems, which was presented at the International Congress of Mathematicians in Paris, 1900, dealing with the representation of a definite form as an SOS. Hilbert also noted that not every positive semidefinite polynomial (or form) is SOS. However, [69] has proved that the numerical examples seem to indicate that the gap between the SOS and nonnegativity polynomial is small. A complete characterization has been outlined by Hilbert in explaining when these two classes are equivalent. There are three cases in which the equality holds: 1. The case of two variables ($n = 2$); 2. The familiar case of quadratic form (i.e., $m = 2$); 3. A surprising case where $P_{3,4} = \sum_{3,4}$; refer [70] for detailed explanations.

Before ending this section, the following lemma is presented, which is useful for our main results later.

Lemma 1. *[52] Let $F(x)$ be an $N \times N$ symmetric polynomial matrix of degree $2d$ in $x \in R^n$. Furthermore, let $Z(x)$ be a column vector whose entries are all monomials in x with a degree no greater than d, and consider the following conditions:*

1. *$F(x) \geq 0$ for all $x \in R^n$;*
2. *$v^T F(x) v$ is an SOS, where $v \in R^N$;*
3. *There exists a positive semidefinite matrix Q such that $v^T F(x) v = (v \otimes Z(x))^T\, Q\, (v \otimes Z(x))$, with $\otimes$ denoting the Kronecker product.*

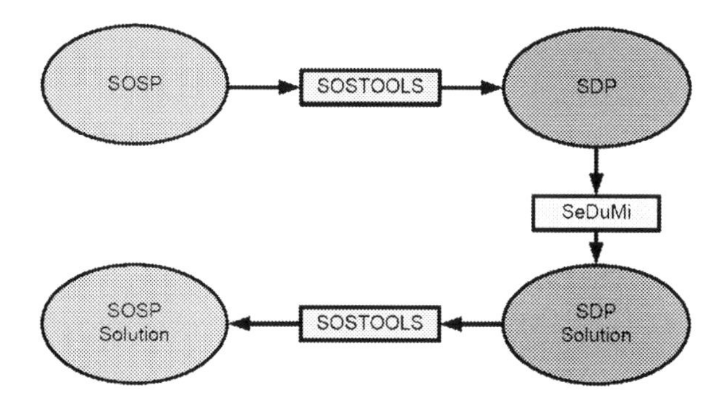

■ **FIGURE 1.2** Diagram depicting how SOS programs (SOSPs) are solved using SOSTOOLS.

It is clear that $F(x)$ being an SOS implies $F(x) \geq 0$, but the converse is generally not true. Furthermore, statement (2) and statement (3) are equivalent.

1.3.2.1 SOSTOOLS

SOSTOOLS is a free, third-party MATLAB toolbox specially designed to handle and solve the SOS programs. The techniques behind it are based on the SOS decomposition for multivariate polynomials, which can be efficiently computed using semidefinite programming frameworks. The availability of SOSTOOLS gives a great advantage to the researchers that involved in SOS polynomial frameworks. Moreover, the SOSTOOLS gives a new direction for solving many hard problems such as global, constrained, and Boolean optimization due to the fact that these technique provide a convex relaxation approach.

The working principles of SOSTOOLS are shown in Fig. 1.2. Basically, the SOSTOOLS automatically convert an SOS program (SOSP) into semidefinite programs (SDPs). Then, it calls the SDP solver and converts the SDP solution back to the solution of the SOS program. In this way the details of the reformulation are abstracted from the user, who can work at the polynomial object level. The user interface of SOSTOOLS has been designed to be simple, easy to use, and transparent while keeping a large degree of flexibility. The current version of SOSTOOLS uses either SeDuMi or SDPT3, both of which are free MATLAB add-ons, as the SDP solver. A detailed description of how SOSTOOLS works can be found in the SOSTOOLS user's guide [77].

Table 1.1 Example of Polynomial Systems.

System	Description
$\dot{x} = x^2$	Finite escape behavior
Lorenz system	Chaos
Brockett integrator	Discontinuous time
Van der Pol system	Limit cycles
Artstein circle	Nonsmooth control
MY conjecture	Global stability

1.4 RESEARCH MOTIVATION

This section provides the reasons that prompted us to conduct research in the framework of controller synthesis for polynomial discrete-time systems. The key motivations for this book come from several sources. The most general motivation comes from the fact that polynomial systems appear in a wide range of applications. This is due to the fact that many nonlinear systems can be modeled as, transformed into, or approximated by polynomial systems. The polynomial systems do not only exist in process control and systems biology, but also appear in many other fields of application, for instance, in mechatronic systems and laser physics; see [78,79]. A few well-known polynomial systems are captured in Table 1.1 (borrowed from [28]). From this table we can see that polynomial systems can show a very rich variety of dynamic behavior. On the other hand, the table also depicts that polynomial systems maybe in general very difficult to study. Therefore, this class of systems is considered in this paper.

The second motivation arises from the fact that state-dependent or parameter-dependent Lyapunov functions ares widely used in the framework of stability analysis and controller design for nonlinear systems. It has been shown recently that state-dependent Lyapunov functions provide a great advantage when dealing with the controller synthesis for polynomial systems [51–54]. This leads to our belief that the utilization of the state-dependent Lyapunov function method should also be effective in designing the controller for polynomial discrete-time systems.

However, with utilization of state-dependent Lyapunov functions, the controller design for polynomial discrete-time systems becomes very difficult. This is due to the fact that the relation between the Lyapunov function and the controller matrix is no longer jointly convex. This problem will be highlighted in detail in Chapter 2. In a continuous-time system, the aforementioned problem can be avoided by assuming that the Lyapunov matrix is only dependent upon the control input whose corresponding rows are zero [53]. Unfortunately, for discrete-time systems, although the same assumption is made, the problem still exists. A possible way to resolve this problem is given in [64], but the results suffer from some conservatism, and such a conservatism has been discussed earlier. In this book, we attempt to relax the problem by incorporating an integrator into the controller structures. In particular, we call this method the integrator method.

Due to the problem discussed before, only a few results are available in the area of controller synthesis in the context of polynomial discrete-time systems [61,64]. As our discussion in the literature shows, the results from both papers suffer from their own conservatism. This consequently motivates us to carry out work on polynomial discrete-time system stabilization. Hence a more general and less conservative result can be provided than the available approaches.

Furthermore, it is also necessary for us to consider the robust controller design for polynomial discrete-time systems because to date, to the authors' knowledge, no results have been presented in this framework that consider the SOS programming technique. For this context, the polytopic uncertainty and norm-bounded uncertainty will be considered because both of them commonly appear in the real world. Besides, the norm-bounded uncertainty is not fully studied in the area of polynomial systems. This is another motivation that leads us to consider the norm-bounded uncertainty in our research work.

The final and somewhat peripheral motivation is that many control design problems are normally formulated in terms of inequalities rather than simple equalities. Moreover, a lot of problems in control engineering can be formulated as polynomial matrix inequalities (PMIs) feasibility problems. Using the SOS decomposition method, such PMIs can be further formulated as SOS constraints [50]. The SOS inequalities framework provides a tractable method to solve the problem through an analytical solution. Furthermore, the advantage of formulating the problem in terms of the SOS inequalities is the availability of toolboxes compatible with MATLAB that are capable for solving the feasibility and optimization problems by interior-point methods. All of these toolboxes are actually SOS-SDP based, and the general concept regarding them has already been explained earlier.

1.5 CONTRIBUTION OF THE BOOK

The focus of this book is to establish novel methodologies for robust stabilization, control with disturbance attenuation, and filter design for a class of polynomial discrete-time systems. The polytopic uncertainties and norm-bounded uncertainties are considered, and the proposed controller is able to handle the appearance of such uncertainties.

The main contribution arises from the incorporation of an integrator into the controller structures. In doing so, a convex solution to the polynomial discrete-time system stabilization with the utilization of a state-dependent Lyapunov function can be obtained in a less conservative way than the available approaches. In light of this integrator method, the problems of robust control and robust H_∞ control for polynomial discrete-time systems are tackled. The integrator method is also applied to the filter design problem.

In this book, we first highlight the problem of the controller design for polynomial discrete-time systems when the state-dependent Lyapunov function is under consideration. Motivated by this problem, we propose a novel method in which an integrator is incorporated into the controller structures. Then, we show that the original systems with the proposed controller can be described in augmented forms. In addition, by choosing the Lyapunov matrix to be only dependent upon the original system's states, a convex solution to the robust control problem and robust H_∞ control problem for polynomial discrete-time systems can be rendered in a less conservative way than available approaches.

In light of the integrator method, we propose a novel methodology for designing a robust nonlinear controller in which the polytopic and norm-bounded uncertainties are under consideration. It should be noted here that, to date, no result is available in this framework that utilizes SOS programming for polynomial discrete-time systems. Furthermore, the interconnection between the nonlinear H_∞ control problem and the robust nonlinear H_∞ control problem is provided through a so-called *scaled* system. This allows us to efficiently solve the robust H_∞ control problem with the existence of norm-bounded uncertainties. Next, we show that by exploiting the integrator method a filter design methodology can also be established for polynomial discrete-time systems. Furthermore, by applying the integrator method the output feedback controller is developed for polynomial discrete-time systems with and without H_∞ performance and also with and without uncertainties.

Motivated by most of the existing control design methods, discrete-time fuzzy polynomial systems cannot guarantee their Lyapunov function to be a radially unbounded polynomial function, and hence the global stability cannot be ensured. This book also provides controller design methods for discrete-time fuzzy polynomial systems that guarantee a radially unbounded polynomial Lyapunov function that ensures the global stability.

Finally, to demonstrate the effectiveness and advantages of the proposed design methodologies of this book, some numerical examples are given. The simulation results also show that the proposed design methodologies can achieve the stability requirement or/and a prescribed performance index.

1.6 BOOK OUTLINE

Chapter 2 describes a nonlinear feedback controller design for polynomial discrete-time systems. In this chapter, the problems of designing a controller for polynomial discrete-time systems are first highlighted. Then, a novel integrator method for solving the problem is proposed. Furthermore, we show that the results can be directly extended to the robust control problem with either polytopic uncertainties or norm-bounded uncertainties. The existence of the proposed controller is given in terms of the solvability of polynomial matrix inequalities (PMIs), which are formulated as SOS constraints and can be solved by the recently developed SOS solvers. The effectiveness of the proposed method is confirmed through simulation examples.

Chapter 3 utilizes the integrator approach proposed in the previous chapter. We provide a less conservative design procedure in the framework of H_∞ control of polynomial discrete-time systems. This result is subsequently extended to the robust H_∞ control problem with the existence of the polytopic uncertainties and norm-bounded uncertainties. The attention here is to design a nonlinear feedback controller such that both stability and a prescribed disturbance attenuation for the closed-loop polynomial discrete-time system are achieved. Based on the SOS-based method, sufficient conditions for the existence of a nonlinear H_∞ controller are given in terms of the solvability of polynomial matrix inequalities (PMIs). Numerical examples are used to demonstrate the effectiveness of the proposed approach.

Chapter 4 deals with the problem of robust nonlinear filtering design for polynomial discrete-time systems. By utilizing the integrator method sufficient conditions for the existence of a robust nonlinear filter are provided in terms of SOS constraints. Numerical examples are given along with the theoretical presentation.

Chapter 5 aims at designing an H_∞ filter for polynomial discrete-time systems with and without uncertainties. The uncertainty under consideration in this chapter is described by integral functional constraints. The objective of an H_∞ nonlinear filtering problem is to design a dynamic nonlinear filter such that the gain from an exogenous input to an estimation error is minimized or guaranteed to be less or equal to a prescribed value for all admissible uncertainties. The effectiveness of the proposed method is confirmed through simulation examples.

Chapter 6 investigates the problem of nonlinear H_∞ static output feedback controller design for polynomial discrete-time systems. In this chapter, we address the problem of H_∞ control in which both stability and a prescribed H_∞ performance are required to be fulfilled. An integrator method is proposed to convert the nonconvex control problem into a convex control problem. Hence, the problem can be solved efficiently by SDP. Sufficient conditions for the existence of the controller are given in terms of the solvability conditions of SOS constraints. It is important to note that the resulting controller gains are in the rational matrix functions of the system output matrices and the additional augmented state. The results are then directly extended to the robust H_∞ output feedback control with polytopic uncertainty and norm-bounded uncertainty.

Most of the existing control design methods of discrete-time fuzzy polynomial systems cannot guarantee their Lyapunov function to be a radially unbounded polynomial function, and hence the global stability cannot be assured. Motivated by this drawback, **Chapter 7** examines the problem of designing a controller for a class of discrete-time nonlinear systems that are represented by discrete-time polynomial fuzzy models. The design methods provided in this chapter guarantee radially unbounded polynomial Lyapunov functions, which ensures global stability. Furthermore, in all the existing methods in the literature, the Lyapunov function is assumed to be a function of the states whose corresponding rows in the control matrix are zero to avoid the nonconvexity problem. This assumption is relaxed in this chapter, where the polynomial Lyapunov function is state dependent without restriction of having zero rows in the control matrix. In addition, a convex solution is obtained by incorporating an integrator into the controller. Sufficient conditions of stability are derived in terms of polynomial matrix inequalities which are solved via SOSTOOLS solvers.

Chapter 8 examines the control problem of uncertain discrete-time polynomial fuzzy systems with H_∞ performance objective using an SOS approach. Uncertainties under consideration are described by norm-bounded constraints. The objective of a nonlinear H_∞ controller is to stabilize the closed-loop system globally and to ensure that the gain from an exogenous input to the regulated output is minimized or guaranteed to be less or equal to a prescribed value for all admissible uncertainties. The design conditions are provided in terms of SOS constraints that can be numerically solved via the SOSTOOLS. The proposed design method guarantees a radially unbounded polynomial Lyapunov function, which ensures global stability.

Concluding remarks are given, and suggestions for future research work are discussed in **Chapter 9**. Finally, some mathematical background knowledge used throughout this research is provided in the **Appendix A**.

REFERENCES

[1] J.K. Åström, B. Wittenmark, Computer Controlled Systems Theory and Design, Prentice Hall Inc., 1984.

[2] K. Ogata, Discrete-Time Control Systems, 2nd edition, Prentice Hall Inc., 1995.

[3] D. Nešić, Dead-Beat Control for Polynomial Systems, PhD dissertation, Australian National University, 1996.

[4] J.J. Slotine, W. Li, Applied Nonlinear Control, Prentice-Hall, Englewood Cliffs, NJ, 1991.

[5] M. Krstic, I. Kanellakopoulos, P.V. Kokopotic, Nonlinear and Adaptive Control Design, John Wiley and Sons, New York, NY, 1995.

[6] P.V. Kokotovic, M. Arcak, Constructive nonlinear control: a historical perspective, Automatica 37 (2001) 637–662.

[7] W.J. Rugh, J.S. Shamma, Research on gain scheduling, Automatica 36 (2000) 1401–1425.

[8] G. Becker, A. Packard, Robust performance of linear parametrically varying systems using parametrically-dependent linear feedback, Systems & Control Letters 23 (1994) 205–215.

[9] F. Wu, X.H. Yang, A. Packard, G. Becker, Induced L_2 norm control for LPV systems with bounded parameter variation rates, International Journal of Robust and Nonlinear Control 6 (1996) 983–998.

[10] P. Apkarian, R. Adams, Advanced gain-scheduling techniques for uncertain systems, IEEE Transactions on Control Systems Technology 6 (1997) 21–32.

[11] J.C. Lo, R. Adams, M.L. Lin, Robust H_∞ nonlinear control via fuzzy static output feedback, IEEE Transactions on Circuit and Systems. I 50 (2003) 1494–1502.

[12] D. Huang, S.K. Nguang, Static output feedback controller design for fuzzy systems: an ILMI approach, Journal of Information Sciences 177 (2007) 3005–3015.

[13] D. Huang, S.K. Nguang, Robust H_∞ static output feedback controller design for fuzzy systems: an ILMI approach, IEEE Transaction on Systems, Man and Cybernetics – Part B: Cybernetics 36 (2006) 216–222.

[14] S.K. Nguang, W. Assawinchaichote, P. Shi, Y. Shi, Robust H_∞ control design for uncertain fuzzy systems with Markovian jumps: an LMI approach, American Control Conference (2005) 1805–1810.

[15] S.K. Nguang, P. Zhang, S.X. Ding, Parity relation based fault estimation for nonlinear systems: an LMI approach, International Journal of Automation and Computing 4 (2) (2007) 164–168.

[16] W. Assawinchaichote, S.K. Nguang, P. Shi, Fuzzy Control and Filter Design for Uncertain Fuzzy Systems, Springer, 2006.

[17] S.K. Nguang, P. Shi, On designing filters for uncertain sampled-data nonlinear systems, Systems & Control Letters 41 (5) (2000) 305–316.

[18] S.K. Nguang, P. Shi, Stabilisation of a class of nonlinear time-delay systems using fuzzy models, in: Proceedings of the 39th IEEE Conference on Decision and Control, 2000, pp. 5–11.

[19] W. Assawinchaichote, S.K. Nguang, P. Shi, E.K. Boukas, H_∞ fuzzy state-feedback control design for nonlinear systems with stability constraints: an LMI approach, Mathematics and Computers in Simulation 78 (4) (2008) 514–531.
[20] S.K. Nguang, Comments on "Robust stabilization of uncertain input-delay systems by sliding mode control with delay compensation", Automatica 37 (10) (2001) 1677.
[21] S.K. Nguang, P. Shi, H_∞ output feedback control of fuzzy system models under sampled measurements, Computers and Mathematics With Applications 46 (5) (2003) 705–717.
[22] J. Zhang, P. Shi, J. Qiu, S.K. Nguang, A novel observer-based output feedback controller design for discrete-time fuzzy systems, IEEE Transactions on Fuzzy Systems 23 (1) (2015) 223–229.
[23] S.K. Nguang, P. Shi, Delay-dependent H_∞ filtering for uncertain time delay nonlinear systems: an LMI approach, IET Control Theory & Applications 1 (1) (2007) 133–140.
[24] Y. Zhang, P. Shi, S.K. Nguang, H.R. Karimi, Observer-based finite-time fuzzy H_∞ control for discrete-time systems with stochastic jumps and time-delays, Signal Processing 97 (2014) 252–261.
[25] S. Chae, S.K. Nguang, SOS based robust H_∞ fuzzy dynamic output feedback control of nonlinear networked control systems, IEEE Transactions on Cybernetics 44 (7) (2014) 1204–1213.
[26] G. Chesi, LMI techniques for optimization over polynomials in control: a survey, IEEE Transaction on Automatic Control 55 (11) (2010) 2500–2510.
[27] J.C. Willems, Dissipative dynamical systems, part I: general theory, Archive for Rational Mechanics and Analysis 45 (1972) 321–351.
[28] C. Ebenbauer, F. Allgower, Analysis and design of polynomial control systems using dissipation inequalities and sum of squares, Journal of Computers and Chemical Engineering 30 (11) (2006) 1601–1614.
[29] R. Mtar, M.M. Belhaouane, M.F. Ghariani, H. Belkhiria, N.B. Braiek, An LMI criterion for the global stability analysis of nonlinear polynomial systems, Nonlinear Dynamics and Systems Theory 9 (2009) 171–183.
[30] M.M. Belhaouane, R. Mtar, H. Belkhiria, N.B. Braiek, Improved results on robust stability analysis and stabilization for a class of uncertain nonlinear systems, International Journal of Computers, Communications & Control 4 (4) (2009).
[31] M.M. Belhaouane, H. Belkhiria, N.B. Braiek, Improved Results on Robust Stability Analysis and Stabilization for a Class of Uncertain Nonlinear Systems, Mathematical Problems in Engineering, Hindawi Publishing, 2010.
[32] D. Cheng, H. Qi, Global stability and stabilization of polynomial systems, in: Proceedings of the 46th Conference on Decision and Control, 2007, pp. 1746–1751.
[33] D. Henrion, J.B. Lasserre, Convergent relaxations of polynomial matrix inequalities and static output feedback, IEEE Transaction on Automatic Control 51 (2) (2006).
[34] J.B. Lasserre, Global optimization with polynomials and the problem of moments, SIAM Journal of Optimal 11 (3) (2001).
[35] L. Vandenberghe, S.P. Boyd, Semidefinite programming, SIAM Review 38 (1) (1996) 49–95.
[36] K. Tanaka, A sum of squares approach to modeling and control of nonlinear dynamical systems with polynomial fuzzy systems, IEEE Transaction on Fuzzy Systems 17 (2009) 911–922.

[37] H.K. Lam, L.D. Seneviratne, Stability analysis of polynomial fuzzy-model-based control systems under perfect/imperfect premise matching, IET Control Theory & Applications 5 (15) (2011).

[38] L. Huang, C.Y. Cheng, G.R. Yu, Design of polynomial controllers for a two-link robot arm using sum-of-squares approachs, in: Proceedings of the 8th Asian Control Conference (ASCC), 2011.

[39] B.W. Sanjaya, B.R. Trilaksono, A. Syaichu-Rohman, Static output feedback control synthesis for nonlinear polynomial fuzzy systems using a sum of squares approach, in: Proceedings of the International Conference on Instrumentation, Communication, Information Technology and Biomedical Engineering, 2011.

[40] A.M. Lyapunov, The General Problem of the Stability of Motion, CRC Press, 1992, English Translation.

[41] F. Rasool, D. Huang, S.K. Nguang, Robust H_∞ output feedback control of networked control systems with multiple quantizers, Journal of the Franklin Institute 349 (3) (2012) 1153–1173.

[42] D. Huang, S.K. Nguang, Robust Control for Uncertain Networked Control Systems With Random Delays, Springer Science & Business Media, 2009.

[43] J. Zhang, A.K. Swain, S.K. Nguang, Robust sensor fault estimation scheme for satellite attitude control systems, Journal of the Franklin Institute 350 (9) (2013) 2581–2604.

[44] Z. Hou, J. Luo, P. Shi, S.K. Nguang, Stochastic stability of Ito differential equations with semi-Markovian jump parameters, IEEE Transactions on Automatic Control 51 (8) (2006) 1383–1387.

[45] F. Rasool, D. Huang, S.K. Nguang, Robust H_∞ output feedback control of discrete-time networked systems with limited information, Systems & Control Letters 60 (10) (2011) 845–853.

[46] S. Saat, S.K. Nguang, Nonlinear H_∞ output feedback control with integrator for polynomial discrete-time systems, International Journal of Robust and Nonlinear Control 25 (2015) 1051–1065.

[47] S. Chae, F. Rasool, S.K. Nguang, A. Swain, Robust mode delay-dependent H_∞ control of discrete-time systems with random communication delays, IET Control Theory & Applications 4 (6) (2010) 936–944.

[48] Y. Zhang, P. Shi, S.K. Nguang, Y. Song, Robust finite-time H_∞ control for uncertain discrete-time singular systems with Markovian jumps, IET Control Theory & Applications 8 (12) (2014) 1105–1111.

[49] S. Boyd, L. Ghaoui, E. Feron, V. Balakrishnan, Linear Matrix Inequalities in System and Control Theory, SIAM, 1994.

[50] P.A. Parrilo, Structured Semidefinite Programs and Semialgebraic Geometry Methods in Robustness and Optimization, PhD dissertation, California Inst. Technol., Pasadena, 2000.

[51] A. Papachristodoulou, S. Prajna, On the construction of Lyapunov functions using the sum of squares decomposition, in: Proceedings of the 41st IEEE Conference on Decision and Control (CDC), Las Vegas, 2002.

[52] S. Prajna, A. Papachristodoulou, F. Wu, Nonlinear control synthesis by sum of squares optimization: a Lyapunov-based approach, in: Proceedings of the 5th Asian Control Conference, 2004, pp. 157–165.

[53] H.J. Ma, G.H. Yang, Fault-tolerant control synthesis for a class of nonlinear systems: sum of squares optimization approach, International Journal of Robust and Nonlinear Control 19 (5) (2009) 591–610.
[54] D. Zhao, J. Wang, An improved H_∞ synthesis for parameter-dependent polynomial nonlinear systems using sos programming, in: American Control Conference, 2009, pp. 796–801.
[55] Q. Zheng, F. Wu, Regional stabilisation of polynomial nonlinear systems using rational Lyapunov functions, International Journal of Control 82 (9) (2009) 1605–1615.
[56] T. Jennawasin, T. Narikiyo, M. Kawanishi, An improved SOS-based stabilization condition for uncertain polynomial systems, in: SICE Annual Conference, 2010, pp. 3030–3034.
[57] D. Zhao, J. Wang, Robust static output feedback design for polynomial nonlinear systems, International Journal of Robust and Nonlinear Control (2009).
[58] S. Saat, M. Krug, S.K. Nguang, A nonlinear static output controller design for polynomial systems: an iterative sums of squares approach, in: 4th International Conference on Control and Mechatronics (ICOM), 2011, pp. 1–6.
[59] S. Saat, M. Krug, S.K. Nguang, Nonlinear H_∞ static output feedback controller design for polynomial systems: an iterative sums of squares approach, in: 6th IEEE Conference on Industrial Electronics and Applications (ICIEA), 2011, pp. 985–990.
[60] S.K. Nguang, S. Saat, M. Krug, Static output feedback controller design for uncertain polynomial systems: an iterative sums of squares approach, IET Control Theory & Applications 5 (9) (2011) 1079–1084.
[61] J. Xu, L. Xie, Y. Wang, Synthesis of discrete-time nonlinear systems: a SOS approach, in: American Control Conference, 2007, pp. 4829–4834.
[62] W. Tan, Nonlinear Control Analysis and Synthesis Using Sums-of-Squares Programming, PhD dissertation, University of California, Berkeley, USA, Spring 2006.
[63] M. Kocvara, M. Stingl, PENBMI User Manual, Tech. Rep., 1.2 edition, Jan. 2004.
[64] H.J. Ma, G.H. Yang, Fault tolerant H_∞ control for a class of nonlinear discrete-time systems: using sum of squares optimization, in: Proceeding of American Control Conference, 2008, pp. 1588–1593.
[65] K. Murty, S.N. Kabadi, Some NP-complete problems in quadratic and nonlinear programming, Mathematical Programming 39 (1987) 117–129.
[66] P. Dorato, W. Yang, C. Abdallah, Robust multiobjective feedback design by quantifier elimination, Journal of Symbolic Computation 24 (1997) 153–159.
[67] M. Jirstrand, Nonlinear control system design by quantifier elimination, Journal of Symbolic Computation 24 (1997) 137–152.
[68] B. Reznick, Uniform denominators in Hilbert's seventeen problem, Mathematische Zeitschrift 220 (1995) 75–97.
[69] P.A. Parrilo, B. Sturmfels, Minimizing polynomial functions, in: Workshop on Algorithm and Quantitative Aspects of Real Algebraic, in: Geometry in Mathematics and Computer Science, 2001.
[70] B. Reznick, Some concrete aspects of Hilbert's 17th problem, in: Contemporary Mathematics, vol. 253, American Mathematics Society, 2005, pp. 251–272.
[71] N.K. Bose, C.C. Li, A quadratic form representation of polynomials of several variables and its application, IEEE Transaction on Automatic Control 14 (1968) 447–448.
[72] Y. Huang, Nonlinear Optimal Control: An Enhanced Quasi-LPV Approach, PhD dissertation, California Inst. Technol., Pasadena, 1998.

[73] S. Prajna, A. Papachristodoulou, P.A. Parrilo, Introducing SOSTOOLS: a general purpose sum of squares programming solver, in: Conference on Decision and Control, vol. 1, 2002, pp. 741–746.
[74] J. Lofberg, YALMIP: a toolbox for modeling and optimization in Matlab, in: IEEE International Symposium on Computer Aided Control Systems Design, 2004, pp. 284–289.
[75] M. Grant, S. Boyd, Graph implementations for nonsmooth convex programs, in: Recent Advances in Learning and Control, in: Lecture Notes in Control and Information Sciences, Springer-Verlag Limited, 2008, pp. 95–110.
[76] D. Henrion, J.B. Lasserre, J. Lofberg, GloptiPoly 3: moments, optimization and semidefinite programming, Optimization Methods and Software 24 (4) (2009) 761–779.
[77] S. Prajna, A. Papachristodoulou, P. Seiler, SOSTOOLS: Sum of Squares Optimization Toolbox for MATLAB, User's Guide, 2004.
[78] S.L. Chen, C.T. Chen, Exact linearization of a voltage-controlled 3-pole active magnetic bearing system, IEEE Transaction on Control Systems Technology 10 (2002) 618–625.
[79] A.N. Pisarchik, B.F. Kuntsevich, Control of multistability in a directly modulated diode laser, IEEE Transaction on Quantum Electronics 38 (12) (2002) 1594–1598.

Chapter 2

Robust nonlinear control for polynomial discrete-time systems

CHAPTER OUTLINE

2.1 INTRODUCTION

The controller design for polynomial discrete-time systems is a hard problem due to the fact that the relation between the Lyapunov function and the controller matrix is always not jointly convex. In continuous-time systems, a convex solution can be achieved by restricting the Lyapunov function to be the only function of states whose corresponding rows in the control matrix are zeroes and whose inverse is of a certain form [1–3]. Unfortunately, this leads to conservative results. In discrete-time systems, the nonconvex problem remains persistent although the same restriction is applied. The attempt to design a state feedback controller for polynomial discrete-time systems can be found in [4]. The proposed methodology suffers from several sources of conservatism, and such conservatisms have been discussed in the preceding chapter.

Motivated by the results in [4] and the problem mentioned, this chapter attempts to convexify the state feedback control problem for polynomial

Analysis and Synthesis of Polynomial Discrete-Time Systems. DOI: 10.1016/B978-0-08-101901-6.00002-5

discrete-time systems in a less conservative way and consequently leads to a less conservative controller design procedure for polynomial discrete-time systems. To be precise, in our work a less conservative design procedure is achieved by incorporating an integrator into the controller structure. In doing so, an original system can be transformed into an augmented system, and the Lyapunov function can be selected to be only dependent upon the original system states. This consequently causes the solution of controller synthesis for polynomial discrete-time systems to be convex and therefore can possibly be solved via SDP. It is important to note here that the resulting controller is given in terms of a rational matrix function of the augmented system. A sufficient condition for the existence of our proposed controller is given in terms of the solvability condition of PMIs, which is formulated as SOS constraints. The problem then can be solved by the recently developed SOS solvers.

On the other hand, one of the important attributes of a good control system design is that closed-loop systems remain stable in the presence of uncertainties [5,6]. Generally, the uncertainty could result from the system simplification or simply from parameter inaccuracies [7]. The existence of such uncertainty could degrade the performance of the system significantly especially for practical systems and may even lead to instability. Hence, it must be handled efficiently to design controllers that operate in a real environment. In this chapter, we attempt to stabilize the polynomial discrete-time system with the existence of norm-bounded uncertainties. The motivation of this chapter arises due to the fact that the norm-bounded uncertainties exist in many real systems, and most uncertain control systems can be approximated by systems with norm-bounded uncertainties. A vast literature is available in the framework of robust control in which the uncertainty is modeled as norm-bounded: see [8–31]. However, none of the listed papers consider the polynomial discrete-time system in their work. Furthermore, to date, to the author's knowledge, no result has been proposed in the framework of robust control problems with norm-bounded uncertainties using the SOS decomposition method.

Therefore, in this chapter, we attempt to design a robust nonlinear feedback controller for an uncertain polynomial discrete-time system with norm-bounded uncertainty. The controller should ensure the uncertain system to be robustly stable. Here, robust stability means that the uncertain system is stable about the origin for all admissible uncertainties. We show that by incorporating the integrator into the controller structure the robust control problem can be converted into a convex solution in a less conservative design procedure. In this work, motivated by the results in [32], the uncertainty is bounded from above. The existence of the proposed robust nonlinear feedback controller is given in terms of the solvability conditions of PMIs, which is formulated as SOS conditions, which then are solved using SOSTOOLS [33].

The rest of this chapter is organized as follows. Section 2.2 provides the main results in which the problem of designing a state feedback controller is highlighted first, and then a novel method is proposed to overcome the problem. The results are then directly extended to the robust control problem with polytopic uncertainty. The validity of our proposed approach is illustrated using numerical examples in Section 2.3. Conclusions are given in Section 2.4.

2.2 MAIN RESULTS

The significance of incorporating the integrator into the controller structure can be seen in this section. We begin this section by presenting the latest results on nonlinear feedback controller design for polynomial discrete-time systems, then followed by our main results on synthesizing the controller without the existence of uncertainties. Finally, these results are subsequently extended to the robust controller design with the existence of polytopic and norm bounded uncertainties.

2.2.1 Nonlinear control for polynomial discrete-time systems

Consider the following dynamic model of a polynomial discrete-time system:

$$x(k+1) = A(x(k))x(k) + B(x(k))u(k), \tag{2.1}$$

where $x(k) \in R^n$ is a state vector, $u(k)$ is an input, and $A(x(k))$ and $B(x(k))$ are polynomial matrices of appropriate dimensions.

For system (2.1), we propose the following state feedback controller:

$$u = K(x(k))x(k). \tag{2.2}$$

For this purpose, we use the standard assumption for the state feedback control where all states vector $x(k)$ are available for feedback. The following theorem is established for system (2.1) with controller (2.2).

Theorem 1. *System (2.1) is asymptotically stable if*

1. *There exist a positive definite symmetric polynomial matrix $P(x(k))$ and a polynomial matrix $K(x(k))$ such that*

$$-(A(x(k)) + B(x(k))K(x(k)))^T P^{-1}(x_+)(A(x(k))$$

$$+ B(x(k))K(x(k))) + P^{-1}(x(k)) > 0, \tag{2.3}$$

or

2. *There exist a positive definite symmetric polynomial matrix $P(x(k))$ and polynomial matrices $K(x(k))$ and $G(x(k))$ such that*

$$\begin{bmatrix} G^T(x(k)) + G(x(k)) - P(x(k)) & * \\ A(x(k))G(x(k)) + B(x(k))K(x(k))G(x(k)) & P(x_+) \end{bmatrix} > 0. \tag{2.4}$$

Proof. Select a Lyapunov function as follows:

$$V(x(k)) = x^T(k)P^{-1}(x(k))x(k). \tag{2.5}$$

Then the difference of (2.5) along (2.1) with (2.2) is given by

$$\begin{aligned} \Delta V(x(k)) &= V(x(k+1)) - V(x(k)) < 0 \\ &= x^T(k+1)P^{-1}(x_+)x(k+1) - x^T(k)P^{-1}(x(k))x(k) \\ &= \left(A(x(k))x(k) + B(x(k))K(x(k))x(k)\right)^T P^{-1}(x_+) \\ &\left(A(x(k))x(k) + B(x(k))K(x(k))x(k)\right) \\ &- x^T(k)P^{-1}(x(k))x(k) \\ &= x^T(k)\left[(A^T(x(k)) + K^T(x(k))B^T(x(k)))P^{-1}(x_+)(A(x(k)) \right. \\ &\left. + B(x(k))K(x(k))) - P^{-1}(x(k))\right]x(k). \end{aligned} \tag{2.6}$$

Now we have to show that (2.4) $\Leftrightarrow$ (2.3). Necessity follows by choosing $G(x(k)) = G^T(x(k)) = P(x(k))$. For sufficiency, suppose that (2.4) holds and thus $G^T(x(k)) + G(x(k)) > P(x(k)) > 0$. This implies that $G(x(k))$ is nonsingular. Since $P(x(k))$ is positive definite, we have the inequality

$$(P(x(k)) - G(x(k)))^T P^{-1}(x(k))(P(x(k)) - G(x(k))) > 0, \tag{2.7}$$

establishing

$$G^T(x(k))P^{-1}(x(k))G(x(k)) > G(x(k)) + G^T(x(k)) - P(x(k)), \tag{2.8}$$

and therefore we have

$$\begin{bmatrix} G^T(x(k))P^{-1}(x(k))G(x(k)) & * \\ A(x(k))G(x(k)) + B(x(k))K(x(k))G(x(k)) & P(x_+) \end{bmatrix} > 0. \tag{2.9}$$

Next, multiplying (2.9) on the right by $diag[G^{-1}(x(k)), I]$ and on the left by $diag[G^{-1}(x(k)), I]^T$, we get

$$\begin{bmatrix} P^{-1}(x(k)) & * \\ A(x(k)) + B(x(k))K(x(k)) & P(x_+) \end{bmatrix} > 0. \tag{2.10}$$

Applying the Schur complement to (2.10), we arrive at (2.3). Knowing that (2.3) holds, we have $\Delta V(x(k)) < 0$ for all $x \neq 0$, which implies that system (2.1) with (2.2) is globally asymptotically stable. This completes the proof. □

Remark 1. The advantages of formulating the problem of the form (2.4) are twofold:

1. The Lyapunov function is decoupled from the system matrices. Therefore, the selection of the polynomial feedback control law can be chosen to be a polynomial of arbitrary degree, which improves the solvability of the nonlinear matrix inequalities by the SOS solver. This also allows the method to be extended to the robust control problem.
2. The number of $P(x(k))$ can be reduced significantly in the problem formulation.

The introduction of this new polynomial matrix method is first proposed in [34] for linear cases and has been adopted by [4] for nonlinear cases. It is also important to note here that this new polynomial matrix $G(x(k))$ is not constrained to be symmetrical.

It is worth mentioning that the conditions given in Theorem 1 are in terms of state-dependent polynomial matrix inequalities (PMIs). Thus, solving this inequality is computationally hard because we need to solve an infinite set of state-dependent PMIs. To relax these conditions, we utilize the SOS decomposition approach as described in [35] and have the following proposition.

Proposition 1. *System (2.1) is asymptotically stable if there exist a symmetric polynomial matrix $P(x(k))$, polynomial matrices $L(x(k))$ and $G(x(k))$, and constants $\epsilon_1 > 0$ and $\epsilon_2 > 0$ such that the following conditions hold for all $x \neq 0$:*

$$v^T[P(x(k)) - \epsilon_1 I]v \quad \text{is an SOS,} \tag{2.11}$$

$$v_1^T\,[M(x(k)) - \epsilon_2 I]\,v_1^T \quad \text{is an SOS,} \tag{2.12}$$

where

$$M(x(k)) = \begin{bmatrix} G^T(x(k)) + G(x(k) - P(x(k)) & * \\ A(x(k))G(x(k)) + B(x(k))L(x(k)) & P(x_+) \end{bmatrix}. \tag{2.13}$$

Meanwhile, v and v_1 are free vectors of appropriate dimensions, and $L(x(k)) = K(x(k))G(x(k))$. Moreover, the nonlinear feedback controller is given by

$$K(x(k)) = L(x(k))G^{-1}(x(k)).$$

Proof. The proof for this proposition can be obtained easily by following the technique given in the proof section of Theorem 1. By Proposition 1, if the inequalities described in (2.11)–(2.12) are feasible, then inequality (2.4) is true. The proof ends. □

Remark 2. Proposition 1 provides a sufficient condition for the existence of a state feedback controller and is given in terms of solutions to a set of parameterized polynomial matrix inequalities (PMIs). Notice that $P(x_+)$ appears in the PMIs, and therefore the inequalities are not convex because $P(x_+) = P(x(k+1)) = P(A(x(k))x(k) + B(x(k))K(x(k))x(k))$. Therefore it is very difficult to directly solve Proposition 1 because the PMIs need to be checked for all combinations of $P(x(k))$ and $K(x(k))$, which results in solving an infinite number of polynomial matrix inequalities. A possible way to resolve this problem has been proposed in [4], where a predefined upper bound is used to limit the effect of the nonconvex term. However, this predefined upper bound is hard to determine beforehand, and the closed-loop stability can only be guaranteed within a bound region. Motivated by this fact, we propose a novel approach in which the aforementioned problem can be removed by incorporating the integrator into the controller structure. This consequently provides a less conservative result on the similar underlying issue. The details of our method are given further.

In [36], the authors assume that $A(x(k))$ can be decomposed as

$$A(x(k)) = E(x(t))Y(x(k)), \tag{2.14}$$

where $Y(x(k))$ and $E(x(k))$ are polynomial matrices. Using this decomposition, they select a state feedback controller of the form

$$K(x(k)) = L(x(k))Y(x(k)), \tag{2.15}$$

where $L(x(k))$ is to be designed.

Let $\mathbb{J} = \{j_1, j_2, \ldots, j_\ell\}$ be the index set where the corresponding j_kth row of $B(x)$ is equal to zero. They define $\tilde{x}(k) = \{x_{j_1}, x_{j_2}, \ldots, x_{j_\ell}\}$ and consider special forms of $P(x(k))$ and $P(x_+)$ that satisfy the following constraints:

$$\begin{aligned} P(x(k)) &= Y^T(x(k))H^T P(\tilde{x}(k))HY(x(k)), && (2.16)\\ P(x_+) &= Y^T(x(k))\hat{H}^T(x(k))P(\tilde{x}_+)\hat{H}(x(k))Y(x(k)), && (2.17) \end{aligned}$$

where H is a constant matrix with full row rank, and $\hat{H}(x(k))$ is a matrix possibly involving the variables of $L(x(k))$. With these choices, the authors have proposed a direct algorithm that combines the advantages of SOS and BMI solvers by incorporating bilinear SOS parameterizations into YALMIP and PENOPT for PENBMI.

Remark 3. Constraints (2.16) and (2.17) require that $HY(x_+)$ can be decomposed as $\hat{H}(x(k))Y(x(k))$. Also note that the choice of $Y(x(k))$ is not unique. Even when $Y(x(k))$ is fixed, the choice of $E(x(k))$ may not be unique. It is obvious that the solvability of the stabilization problem and system performance may depend on the choice of these matrices. However, a limitation is that there is no systematic procedure to generate them so far.

Example 1. Consider the two-dimensional system

$$\begin{bmatrix} x_1(k+1) \\ x_2(k+1) \end{bmatrix} = \begin{bmatrix} -x_1^2(k) + x_2(k) \\ x_1(k) \end{bmatrix} + \begin{bmatrix} x_1^2 \\ 0 \end{bmatrix} u. \tag{2.18}$$

We can choose

$$Y(x) = \begin{bmatrix} 1 & 0 \\ 0 & 1 \end{bmatrix}; \; E(x) = \begin{bmatrix} -x_1 & 1 \\ 1 & 0 \end{bmatrix}; \; P(x) = Y^T(x)H^T P(x_2)HY(x);$$

$$P(x_+) = Y^T(x)\hat{H}^T(x(k))P(x_{2+})\hat{H}(x(k))Y(x(k)) \text{ and } \hat{H}(x(k)) = H.$$

In [4], a special form of one-degree polynomial matrix $P(x(k))$ whose (i, j)th entry is given by

$$p_{ij}(x(k)) = p_{ij}^{(0)} + \sum_{\ell=1}^{n} p_{ij}^{(\ell)} x_\ell = P_{ij}^{(0)} + [p_{ij}^{(1)}, \ldots, p_{ij}^{(n)}]x(k), \tag{2.19}$$

where $p_{ij}^{(\ell)}, i = 1, \ldots, n, \; j = 1, \ldots, n, \; \ell = 1, \ldots, n$, are scalars. To convert the nonjointly convex problem into a semidefinite programming, they define the polynomial matrix $\tilde{P}(x_+)$ with (i, j)th entry

$$\tilde{p}_{ij} = \tilde{p}_{ij}^{(0)} + [p_{ij}^{(1)}, \ldots, p_{ij}^{(n)}]A(x(k))x(k) \tag{2.20}$$

and zero optimization of μ subject to

$$\begin{bmatrix} \mu & [p_{ij}^{(1)}, \ldots, p_{ij}^{(n)}]B(x(k)) \\ * & I \end{bmatrix} \geq 0. \tag{2.21}$$

If the minimum of μ is zero, then $[p_{ij}^{(1)}, \ldots, p_{ij}^{(n)}]B(x(k))$, which makes the nonjointly convex term, $[p_{ij}^{(1)}, \ldots, p_{ij}^{(n)}]B(x(k))K(x(k))x(k)$ zero.

Remark 4. The drawbacks of this approach are that $P(x(k))$ must be a first-degree polynomial matrix and the zero optimization of μ is hard to determine before hand, that is, if μ is not equal to zero, then this design method can only ensure a local stability.

2.2.1.1 The integrator approach

In this section, we show that by incorporating the integrator into the controller structure the controller synthesis for polynomial discrete-time systems can be convexified and a less restrictive design procedure can be achieved.

We propose the following nonlinear feedback controller with integrator:

$$\begin{cases} x_c(k+1) = x_c(k) + A_c(x, x_c), \\ u(k) = x_c(k), \end{cases} \tag{2.22}$$

where x_c is an additional state or controller state, $A_c(x, x_c)$ is an input function of the integrator, and $u(k)$ is an input to the system. Here, the objective is to stabilize system (2.1) with controller (2.22).

System (2.1) with controller (2.22) can be described as follows:

$$\hat{x}(k+1) = \hat{A}(\hat{x}(k))\hat{x}(k) + \hat{B}(\hat{x}(k))A_c(x, x_c), \tag{2.23}$$

where

$$\hat{A}(\hat{x}(k)) = \begin{bmatrix} A(x(k)) & B(x(k)) \\ 0 & 1 \end{bmatrix}; \quad \hat{B}(\hat{x}(k)) = \begin{bmatrix} 0 \\ 1 \end{bmatrix}; \quad \hat{x}(k) = \begin{bmatrix} x(k) \\ x_c(k) \end{bmatrix}; \tag{2.24}$$

Next, we assume $A_c(x, x_c)$ to be of the form $A_c(x, x_c) = \hat{A}_c(\hat{x}(k))\hat{x}(k)$. Therefore (2.23) can be re-written as follows:

$$\hat{x}(k+1) = \hat{A}(\hat{x}(k))\hat{x}(k) + \hat{B}(\hat{x}(k))\hat{A}_c(\hat{x}(k))\hat{x}(k) \tag{2.25}$$

where $\hat{A}(\hat{x}(k))$ and $\hat{B}(\hat{x}(k))$ are as described in (2.24). Meanwhile, $\hat{A}_c(\hat{x}(k))$ is a $1 \times (n+1)$ polynomial matrix, where n is the number of original states.

Remark 5. The above idea of introducing an additional dynamic is not new; see [37,38]. However, the authors in [37,38] used this method to overcome the problem of designing a robust controller for linear systems with norm-bounded uncertainties. In contrast, we propose this method to convexify the controller synthesis problem for polynomial discrete-time systems. Generally, the method corresponds to dynamic state feedback rather than the static state feedback. Notice that using this method, a simple form for the $\hat{B}(\hat{x}(k))$ can be obtained, and it is always in the form of $[0, 1]^T$.

Sufficient conditions for the existence of our proposed controller are given in the following theorem.

Theorem 2. *System (2.1) is stabilizable via the nonlinear feedback controller of the form (2.22) if there exist a symmetric polynomial matrix $\hat{P}(x(k))$ and polynomial matrices $\hat{L}(\hat{x}(k))$ and $\hat{G}(\hat{x}(k))$ such that the following conditions are satisfied for all $x \neq 0$:*

$$\hat{P}(x(k)) > 0, \tag{2.26}$$

$$M_1(\hat{x}(k)) > 0, \tag{2.27}$$

where

$$M_1(\hat{x}(k)) = \begin{bmatrix} \hat{G}^T(\hat{x}(k)) + \hat{G}(\hat{x}(k)) - \hat{P}(x(k)) & * \\ \hat{A}(\hat{x}(k))\hat{G}(\hat{x}(k)) + \hat{B}(\hat{x}(k))\hat{L}(\hat{x}(k)) & \hat{P}(x_+) \end{bmatrix}. \tag{2.28}$$

Moreover, the nonlinear feedback controller is given by

$$\begin{aligned} x_c(k+1) =& x_c(k) + A_c(x, x_c), \\ u(k) =& x_c(k), \end{aligned}$$

where $A_c(x, x_c) = \hat{A}_c(\hat{x}(k))\hat{x}(k)$ with $\hat{A}_c(\hat{x}(k)) = \hat{L}(\hat{x}(k))\hat{G}^{-1}(\hat{x}(k))$.

Proof. The Lyapunov function is chosen as follows:

$$\hat{V}(\hat{x}(k)) = \hat{x}^T(k)\hat{P}^{-1}(x(k))\hat{x}(k). \tag{2.29}$$

Note that the Lyapunov function (2.29) is such that the polynomial matrix $\hat{P}(x(k))$ depends only upon the original system states $x(k)$. By employing this kind of representation, we can avoid the existence of nonconvex terms in our formulation. Based on the Lyapunov stability theory [39], the closed-loop system (2.1) with (2.22) is stable if there exists a Lyapunov function (2.29) > 0 such that

$$\begin{aligned} &\Delta\hat{V}(\hat{x}(k)) = \hat{x}(k+1)^T\hat{P}^{-1}(x_+)\hat{x}(k+1) - \hat{x}^T(k)\hat{P}^{-1}(x(k))\hat{x}(k) \\ &= \big(\hat{A}(\hat{x}(k))\hat{x}(k) + \hat{B}(\hat{x}(k))\hat{A}_c(\hat{x}(k))\hat{x}(k)\big)^T\hat{P}^{-1}(x_+)\big(\hat{A}(\hat{x}(k))\hat{x}(k) \\ &+ \hat{B}(\hat{x}(k))\hat{A}_c(\hat{x}(k))\hat{x}(k)\big) - \hat{x}^T(k)\hat{P}^{-1}(x(k))\hat{x}(k) \\ &= \hat{x}^T(k)\big[\big(\hat{A}^T(\hat{x}(k)) + \hat{A}_c^T(\hat{x}(k))\hat{B}^T(\hat{x}(k))\big)\hat{P}^{-1}(x_+)\big(\hat{A}(\hat{x}(k)) \\ &+ \hat{B}(\hat{x}(k))\hat{A}_c(\hat{x}(k))\big) - \hat{P}^{-1}(x(k))\big]\hat{x}(k) < 0. \end{aligned} \tag{2.30}$$

Now, suppose that (2.27) is feasible. Then applying the Schur complement to it, we can easily arrive at

$$\left[\left(\hat{A}^T(\hat{x}) + \hat{A}_c^T(\hat{x})\hat{B}^T(\hat{x})\right)\hat{P}^{-1}(x_+)\left(\hat{A}(\hat{x}) + \hat{B}(\hat{x})\hat{A}_c(\hat{x}) - \hat{P}^{-1}(x)\right] < 0. \tag{2.31}$$

Knowing that (2.31) holds, we have $\Delta\hat{V}(\hat{x}(k)) < 0$, which implies that system (2.1) with controller (2.22) is globally asymptotically stable. □

Again, Theorem 2 is computationally hard because it requires solving an infinite set of state-dependant PMIs. To relax this problem, we utilize the SOS decomposition based the SDP method and have the following corollary.

Corollary 1. *System* (2.1) *is stabilizable via the nonlinear feedback controller of the form* (2.22) *if there exist a symmetric polynomial matrix* $\hat{P}(x(k))$, *polynomial matrices* $\hat{L}(\hat{x}(k))$ *and* $\hat{G}(\hat{x}(k))$, *and constants* $\epsilon_1 > 0$ *and* $\epsilon_2 > 0$ *such that the following conditions are satisfied for all* $x \neq 0$*:*

$$v_3^T[\hat{P}(x(k)) - \epsilon_1 I]v_3 \quad \text{is an SOS,} \tag{2.32}$$

$$v_4^T\left[M_1(\hat{x}(k)) - \epsilon_2 I\right]v_4 \quad \text{is an SOS,} \tag{2.33}$$

where

$$M_1(\hat{x}(k)) = \begin{bmatrix} \hat{G}^T(\hat{x}(k)) + \hat{G}(\hat{x}(k)) - \hat{P}(x(k)) & * \\ \hat{A}(\hat{x}(k))\hat{G}(\hat{x}(k)) + \hat{B}(\hat{x}(k))\hat{L}(\hat{x}(k)) & \hat{P}(x_+) \end{bmatrix}, \tag{2.34}$$

and v_3 *and* v_4 *are free vectors of appropriate dimensions. Moreover, the nonlinear feedback controller is given by*

$$x_c(k+1) = x_c(k) + A_c(x, x_c),$$
$$u(k) = x_c(k),$$

where $A_c(x, x_c) = \hat{A}_c(\hat{x}(k))\hat{x}(k)$ *with* $\hat{A}_c(\hat{x}(k)) = \hat{L}(\hat{x}(k))\hat{G}^{-1}(\hat{x}(k))$.

Proof. The proof follows directly from the combination of the proofs given in Theorem 2 and Proposition 1. □

Remark 6. The advantages of formulating the problem in the form of Corollary 1 are twofold:

1. The solution given by Corollary 1 is convex and hence allows the problem to be solved computationally via SDP. This is true because the term in $\hat{P}(x_+)$ is now jointly convex. To prove this, refer to the

Lyapunov function (2.29) where the state-dependent Lyapunov matrix $\hat{P}(x(k))$ depends only upon the original system states. Therefore, we have $\hat{P}(x_+) = P(A(x(k))x(k) + B(x(k))x_c(k)$, where $x_c(k)$ is an additional or controller state. This consequently makes the terms in $\hat{P}(x(k))$ jointly convex. We can see that using this integrator method, the nonconvex term is not required to be bounded by the upper bound value. Therefore, our result holds globally. In contrast, to achieve a convex solution to the controller synthesis of polynomial discrete-time systems [4], the nonconvex term must be bounded by a predefined upper bound value, and hence their results are local. Based on the mentioned reason, our method provides a design procedure for controller synthesis of polynomial discrete-time systems less conservative than in [4].

2. In [36], to render a convex solution to the controller design problem, the Lyapunov function must be selected such that its matrix only depends upon the input matrix whose corresponding rows are zeros. In contrast, using our proposed method, no assumption is required to achieve a convex solution to the controller design problem of a polynomial discrete-time system. This is because the original system input matrix $B(x(k))$ is governed in the system matrices of augmented system (2.23) and the Lyapunov function now depends upon the $\hat{B}(x(k))$, which is always $[0, 1]^T$.

2.2.2 Robust nonlinear feedback control design for polynomial discrete-time systems

In this subsection, we begin by considering polynomial discrete-time systems with parametric uncertainties, then followed by systems with norm-bounded uncertainties.

With parametric uncertainties

Consider the following polynomial discrete-time system with parametric uncertainties:

$$x(k+1) = A(x(k), \theta)x(k) + B(x(k), \theta)u(k), \tag{2.35}$$

where the matrices $\cdot(x(k), \theta)$ are defined as

$$\begin{aligned} A(x(k), \theta) &= \sum_{i=1}^{q} A_i(x(k))\theta_i, \\ B(x(k), \theta) &= \sum_{i=1}^{q} B_i(x(k))\theta_i. \end{aligned} \tag{2.36}$$

The vector of constant uncertainty $\theta = \left[\theta_1, \ldots, \theta_q\right]^T \in \mathbb{R}^q$ satisfies

$$\theta \in \Theta \triangleq \left\{\theta \in \mathbb{R}^q : \theta_i \geq 0, i = 1, \ldots, q, \sum_{i=1}^{q} \theta_i = 1\right\}. \tag{2.37}$$

A robust nonlinear feedback controller is given as

$$\begin{cases} x_c(k+1) = x_c(k) + A_c(x, x_c), \\ u(k) = x_c(k), \end{cases} \tag{2.38}$$

where x_c is the controller state variable, and $A_c(x, x_c)$ is a design input function. The objective of this controller is to robustly stabilize the uncertain system (2.35).

With controller (2.38), we have the system

$$\hat{x}(k+1) = \hat{A}(\hat{x}(k), \theta)\hat{x}(k) + \hat{B}(\hat{x}(k), \theta)u(k), \tag{2.39}$$

where

$$\hat{A}(\hat{x}(k), \theta) = \sum_{i=1}^{q} \hat{A}_i(\hat{x}(k))\theta_i; \quad \hat{B}(\hat{x}(k), \theta) = \sum_{i=1}^{q} \hat{B}_i(\hat{x}(k))\theta_i \tag{2.40}$$

with

$$\hat{A}(\hat{x}(k)) = \begin{bmatrix} A(x(k)) & B(x(k)) \\ 0 & 1 \end{bmatrix}; \quad \hat{B}(\hat{x}(k)) = \begin{bmatrix} 0 \\ 1 \end{bmatrix}; \quad \hat{x}(k) = \begin{bmatrix} x(k) \\ x_c(k) \end{bmatrix}. \tag{2.41}$$

We further define the parameter-dependent Lyapunov function

$$\tilde{V}(x(k)) = \hat{x}^T(k) \left(\sum_{i=1}^{q} \hat{P}_i(x(k))\theta_i\right)^{-1} \hat{x}(k). \tag{2.42}$$

From the integral approach given in previous section we have the following theorem:

Theorem 3. *Given constants $\epsilon_1 > 0$ and $\epsilon_2 > 0$, system* (2.35) *with the nonlinear feedback controller* (2.38) *is robustly stable for $x \neq 0$ and $i = 1, \ldots, q$ if there exist common polynomial matrices $\hat{G}(\hat{x}(k))$ and $\hat{L}(\hat{x}(k))$ and a symmetric polynomial matrix $\hat{P}_i(x(k))$ such that the following conditions hold for all $x \neq 0$:*

$$v_5^T[\hat{P}_i(x(k)) - \epsilon_1 I]v_5 \quad \textit{is an SOS}, \tag{2.43}$$

$$v_6^T\left[M_2(\hat{x}(k)) - \epsilon_2 I\right]v_6 \quad \textit{is an SOS}, \tag{2.44}$$

where

$$M_2(\hat{x}(k)) = \begin{bmatrix} \hat{G}^T(\hat{x}(k)) + \hat{G}(\hat{x}(k)) - \hat{P}_i(x(k)) & * \\ \hat{A}_i(\hat{x}(k))\hat{G}(\hat{x}(k)) + \hat{B}_i(\hat{x}(k))\hat{L}(\hat{x}(k)) & \hat{P}_i(x_+) \end{bmatrix}. \tag{2.45}$$

Moreover, the robust nonlinear feedback is given as

$$\begin{aligned} x_c(k+1) &= x_c(k) + A_c(x, x_c), \\ u(k) &= x_c(k), \end{aligned}$$

where $A_c(x, x_c) = \hat{A}_c(\hat{x}(k))\hat{x}(k)$ *with* $\hat{A}_c(\hat{x}(k)) = \hat{L}(\hat{x}(k))\hat{G}^{-1}(\hat{x}(k))$.

Proof. The proof follows directly as a convex combination of several systems of the form (2.35) for common (2.38). □

With norm-bounded uncertainties

Consider the following dynamic model of an uncertain polynomial discrete-time system:

$$\begin{aligned} x(k+1) = {} & A(x(k))x(k) + \Delta A(x(k))x(k) + B_u(x(k))u(k) \\ & + \Delta B_u(x(k))u(k), \end{aligned} \tag{2.46}$$

where $x(k) \in R^n$ is the state vector, $u(k) \in R^m$ is the input, and $A(x(k))$ and $B_u(x(k))$ are polynomial matrices of appropriate dimensions. Meanwhile, $\Delta A(x(k))$ and $\Delta B_u(x(k))$ represent the uncertainties in the system and satisfy the following assumption.

Assumption. The admissible parameter uncertainties considered here are assumed to be norm-bounded and described by the following form:

$$\begin{bmatrix} \Delta A(x(k)) & \Delta B_u(x(k)) \end{bmatrix} = H(x(k))F(x(k)) \begin{bmatrix} E_1(x(k)) & E_2(x(k)) \end{bmatrix}, \tag{2.47}$$

where $H(x(k))$, $E_1(x(k))$, and $E_2(x(k))$ are known polynomial matrices of appropriate dimensions, and $F(x(k))$ is an unknown state-dependent matrix function satisfying

$$\left\| F^T(x(k))F(x(k)) \right\| \leq I. \tag{2.48}$$

□

To avoid the nonconvexity in $P(x(k+1))$, we propose the following nonlinear controller:

$$\begin{cases} x_c(k+1) = x_c(k) + A_c(x, x_c), \\ u(k) = x_c(k), \end{cases} \tag{2.49}$$

where $A_c(x, x_c)$ is the input function of the integrator, x_c is the controller state, and $u(k)$ is the input to the system. It has been shown in previous section that by selecting the controller of the form (2.49), a convex solution to $P(x(k+1))$ can be rendered efficiently.

Next, we assume that $A_c(x, x_c) = \hat{A}_c(\hat{x}(k))\hat{x}(k)$, where $\hat{A}_c(\hat{x}(k))$ is a new polynomial matrix of dimension $1 \times (n+1)$; n is the number of original states. Now, system (2.46) with controller (2.49) can be written as follows:

$$\hat{x}(k+1) = \hat{A}(\hat{x}(k))\hat{x}(k) + \Delta\hat{A}(\hat{x}(k))\hat{x}(k) + \hat{B}_u(\hat{x}(k))\hat{A}_c(\hat{x}(k))\hat{x}(k), \tag{2.50}$$

where

$$\begin{aligned} \hat{A}(\hat{x}(k)) &= \begin{bmatrix} A(x(k)) & B(x(k)) \\ 0 & 1 \end{bmatrix}; \quad \hat{B}_u(\hat{x}(k)) = \begin{bmatrix} 0 \\ 1 \end{bmatrix}; \\ \Delta\hat{A}(\hat{x}(k)) &= \hat{H}(\hat{x}(k))\hat{F}(\hat{x}(k))\hat{E}(\hat{x}(k)) \\ &= \begin{bmatrix} H(x(k)) \\ 0 \end{bmatrix} F(x(k)) \begin{bmatrix} E_1(x(k)) & E_2(x(k)) \end{bmatrix}; \\ \hat{x} &= \begin{bmatrix} x(k) \\ x_c(k) \end{bmatrix}. \end{aligned} \tag{2.51}$$

The objective here is to design a nonlinear feedback controller of the form (2.49) such that system (2.46) with (2.49) is robustly stable. Here, robust stability means that the uncertain system (2.46) is asymptotically stable about the origin for all admissible uncertainties.

Theorem 4. *System* (2.46) *is stabilizable via the nonlinear feedback control of the form* (2.49) *if there exist a symmetric polynomial matrix* $\hat{P}(x(k))$, *polynomial matrices* $\hat{L}(\hat{x}(k))$ *and* $\hat{G}(\hat{x}(k))$, *and a polynomial function* $\epsilon(\hat{x}(k) > 0$ *such that the following conditions hold for all* $x \neq 0$*:*

$$\hat{P}(x(k)) > 0, \tag{2.52}$$

$$\begin{bmatrix} -(\hat{G}^T(\hat{x}(k)) + \hat{G}(\hat{x}(k)) \\ \quad -\hat{P}(x(k))) & * & * \\ \hat{E}(\hat{x}(k))\hat{G}(\hat{x}(k)) & -\epsilon(\hat{x}(k))I & * \\ \hat{A}(\hat{x}(k))\hat{G}(x(k)) \\ \quad +\hat{B}_u(\hat{x}(k))\hat{L}(\hat{x}(k)) & 0 & -P(x_+) + \epsilon(\hat{x}(k))\hat{H}(\hat{x}(k))\hat{H}^T(\hat{x}(k)) \end{bmatrix} < 0. \tag{2.53}$$

Moreover, the nonlinear controller is given by

$$x_c(k+1) = x_c(k) + A_c(x, x_c),$$

$$u(k) = x_c(k),$$

where $A_c(x, x_c) = \hat{A}_c(\hat{x}(k))\hat{x}(k)$ *with* $\hat{A}_c(\hat{x}(k)) = \hat{L}(\hat{x}(k))\hat{G}^{-1}(\hat{x}(k))$.

Proof. Select a state-dependent Lyapunov function of the form

$$\hat{V}(\hat{x}(k)) = \hat{x}^T(k)\hat{P}^{-1}(x(k))\hat{x}(k). \tag{2.54}$$

The difference between $\hat{V}(x(k+1)$ and $\hat{V}(x(k))$ of (2.54) along (2.50) is given by

$$\begin{aligned}\Delta\hat{V}(x(k)) &= \hat{V}(x(k+1) - \hat{V}(x(k))\\ &= \hat{x}^T(k+1)\hat{P}^{-1}(x(k+1))\hat{x}(k+1) - \hat{x}^T(k)\hat{P}^{-1}(x(k))\hat{x}(k)\\ &= \hat{x}^T(k)\big[\Omega(\hat{x}(k))\big]\hat{x}(k),\end{aligned} \tag{2.55}$$

where

$$\begin{aligned}\Omega(\hat{x}(k)) = {} & \big(\hat{A}(\hat{x}(k)) + \hat{B}_u(\hat{x}(k))\hat{A}_c(\hat{x}(k))\\ & + \hat{H}(\hat{x}(k))\hat{F}(\hat{x}(k))\hat{E}(\hat{x}(k))\big)^T\hat{P}^{-1}(x_+)\big(\hat{A}(\hat{x}(k))\\ & + \hat{B}_u(\hat{x}(k))\hat{A}_c(\hat{x}(k)) + \hat{H}(\hat{x}(k))\hat{F}(\hat{x}(k))\hat{E}(\hat{x}(k))\big) - \hat{P}^{-1}(x(k)).\end{aligned} \tag{2.56}$$

Next, we have to show that $\Omega(\hat{x}(k)) < 0$. To show this, suppose that inequality (2.53) holds. Then, from the block (1, 1) of (2.53) we have $\hat{G}^T(\hat{x}(k)) + \hat{G}(\hat{x}(k)) > \hat{P}(x(k)) > 0$. This implies that $\hat{G}(\hat{x}(k))$ is nonsingular, and since $\hat{P}(x(k))$ is positive definite, we have

$$\big(\hat{P}(x(k)) - \hat{G}(\hat{x}(k))\big)^T\hat{P}^{-1}(x(k))\big(\hat{P}(x(k)) - \hat{G}(\hat{x}(k))\big) > 0, \tag{2.57}$$

which yields

$$\hat{G}^T(\hat{x}(k))\hat{P}^{-1}(x_+)\hat{G}(\hat{x}(k)) \geq \hat{G}^T(\hat{x}(k)) + \hat{G}(\hat{x}(k)) - \hat{P}(x(k)). \tag{2.58}$$

This immediately gives

$$\begin{bmatrix} \begin{array}{c}-\hat{G}^T(\hat{x}(k))\hat{P}^{-1}(x(k)))\\ \times(\hat{G}(\hat{x}(k))\end{array} & * & * \\ \hat{E}(\hat{x}(k))\hat{G}(\hat{x}(k)) & -\epsilon(\hat{x}(k))I & * \\ \begin{array}{c}\hat{A}(\hat{x}(k))\hat{G}(x(k))\\ +\hat{B}_u(\hat{x}(k))\hat{L}(\hat{x}(k))\end{array} & 0 & -P(x_+) + \epsilon(\hat{x}(k))\hat{H}(\hat{x}(k))\hat{H}^T(\hat{x}(k)) \end{bmatrix} < 0. \tag{2.59}$$

Then, multiplying on the right of (2.59) by $diag[G^{-1}(\hat{x}(k)), I, I]$ and on the left by $diag[G^{-1}(\hat{x}(k)), I, I]^T$ and knowing that $\hat{L}(\hat{x}(k)) = \hat{A}_c(\hat{x}(k))\hat{G}(\hat{x}(k))$, we arrive at

$$\begin{bmatrix} -\hat{P}^{-1}(x) & * & * \\ \hat{E}(\hat{x}) & -\epsilon(\hat{x})I & * \\ \hat{A}(\hat{x}) + \hat{B}_u(\hat{x})\hat{A}_c(\hat{x}) & 0 & -P(x_+) + \epsilon(\hat{x})\hat{H}(\hat{x})\hat{H}^T(\hat{x}) \end{bmatrix} < 0. \tag{2.60}$$

Similarly,

$$\begin{bmatrix} -\hat{P}^{-1}(x) + \frac{1}{\epsilon(\hat{x})}\hat{E}^T(\hat{x})\hat{E}(\hat{x}) & * \\ \hat{A}(\hat{x}) + \hat{B}_u(\hat{x})\hat{A}_c(\hat{x}) & -P(x_+) + \epsilon(\hat{x})\hat{H}(\hat{x})\hat{H}^T(\hat{x}) \end{bmatrix} < 0. \tag{2.61}$$

Furthermore, knowing that

$$0 \leq x(k)^T \begin{bmatrix} \frac{\hat{E}^T(x(k))}{\epsilon(\hat{x}(k))^{1/2}} \\ -\epsilon(\hat{x}(k))^{1/2}\hat{H}(x(k))\hat{F}(x(k)) \end{bmatrix} \times \begin{bmatrix} \frac{\hat{E}(x(k))}{\epsilon(\hat{x}(k))^{1/2}} & -\epsilon(\hat{x}(k))^{1/2}\hat{F}^T(x(k))\hat{H}^T(x(k)) \end{bmatrix} x(k), \tag{2.62}$$

we have

$$\begin{bmatrix} \frac{1}{\epsilon(\hat{x}(k))}\hat{E}^T(x(k))\hat{E}(x(k)) & 0 \\ 0 & \epsilon(\hat{x}(k))\hat{H}(x(k))\hat{H}^T(x(k)) \end{bmatrix} \geq \begin{bmatrix} 0 & \hat{E}^T(x(k))\hat{F}^T(x(k))\hat{H}^T(x(k)) \\ \hat{H}(x(k))\hat{F}(x(k))\hat{E}(x(k)) & 0 \end{bmatrix}. \tag{2.63}$$

Therefore, using (2.63), Eq. (2.61) can now be written as

$$\begin{bmatrix} -\hat{P}^{-1}(x(k))) & * \\ \hat{A}(\hat{x}(k)) + \hat{B}_u(\hat{x}(k))\hat{A}_c(\hat{x}(k)) + \hat{H}(\hat{x}(k))\hat{F}(\hat{x}(k))\hat{E}(\hat{x}(k)) & -P(x_+) \end{bmatrix} < 0. \tag{2.64}$$

Then, utilizing the Schur complement to (2.64) results in

$$\Omega(\hat{x}(k)) < 0, \tag{2.65}$$

where $\Omega(\hat{x}(k))$ is as described in (2.56). Knowing that (2.65) holds, then we have $\Delta\hat{V}(x(k)) < 0$ for all $x \neq 0$, which implies that system (2.46) with (2.49) is robustly stable. □

Unfortunately, it is hard to solve Theorem 4 because it is given in terms of state-dependent PMIs. Solving this inequality is computationally hard because it requires solving an infinite set of PLMIs. The SOS-based SDP method can provide a computational relaxation for the sufficient condition of Theorem 4. Therefore, the modified SOS-based conditions of Theorem 4 are given as follows.

Proposition 2. *System (2.50) is stabilizable via the nonlinear feedback control of the form (2.49) if there exist a symmetric polynomial matrix $\hat{P}(x(k))$, polynomial matrices $\hat{L}(\hat{x}(k))$ and $\hat{G}(\hat{x}(k))$, a polynomial function $\epsilon(\hat{x}(k)) > 0$, and constants $\epsilon_1 > 0$ and $\epsilon_2 > 0$ such that the following conditions hold for all $x \neq 0$:*

$$v_1^T\left[\hat{P}(x(k)) - \epsilon_1 I\right]v_1 \quad \text{is an SOS,} \tag{2.66}$$

$$-v_2^T\left[M(\hat{x}(k)) + \epsilon_2 I\right]v_2 \quad \text{is an SOS,} \tag{2.67}$$

where

$$M(\hat{x}(k)) =$$
$$\begin{bmatrix} -(\hat{G}(\hat{x}(k)) + \hat{G}^T(\hat{x}(k)) - \hat{P}(x(k))) & * & * \\ \hat{E}(\hat{x}(k))\hat{G}(\hat{x}(k)) & -\epsilon(\hat{x}(k))I & * \\ \hat{A}(\hat{x}(k))\hat{G}(x(k)) + \hat{B}_u(\hat{x}(k))\hat{L}(\hat{x}(k)) & 0 & -P(x_+) + \epsilon(\hat{x}(k))\hat{H}(\hat{x}(k))\hat{H}^T(\hat{x}(k)) \end{bmatrix}, \tag{2.68}$$

and v_1 and v_2 are free vectors of appropriate dimensions. Moreover, the nonlinear controller is given by

$$\begin{aligned} x_c(k+1) &= x_c(k) + A_c(x, x_c), \\ u(k) &= x_c(k), \end{aligned}$$

where $A_c(x, x_c) = \hat{A}_c(\hat{x}(k))\hat{x}(k)$ with $\hat{A}_c(\hat{x}(k)) = \hat{L}(\hat{x}(k))\hat{G}^{-1}(\hat{x}(k))$.

Proof. This can be carried out via similar technique to the proof shown in Theorem 4.

Remark 7. It is always necessary to include the SOS constrain $\epsilon_1 > 0$ to guarantee the negative definiteness of inequalities (2.66)–(2.67) [40].

Remark 8. The drawback of [36] is that the original systems must be transformed or decomposed into certain forms; however, those decompositions are hard to determine, and they are not unique. The solvability of their stabilization problem depends on the choice of the decomposition matrices and the transformation matrix. Furthermore, to render a convex solution to the controller synthesis problem, the Lyapunov function must be of a special form and dependent upon the zeros row of the control matrix. If there are no zero rows in the control matrix, then they need to transform their original systems to introduce as many rows as possible in their control matrix, which often leads to nonlinear descriptor systems. As stated in their paper, there is no systematic procedure to generate them till now. In [4], a predefined upper bound on the nonconvex term needs to be determined using an optimization approach. If the optimization fails to give a feasible solution to that problem, then the overall algorithm will fail to provide a feasible solution too. The results obtained in [4] are local results, and the Lyapunov function can only be a function of states whose corresponding rows in control matrix are zeros. In summary, our method may yield less conservative results than [36, 4] because the nonconvex term does not require to be bounded, our results are global, and the Lyapunov function does not depend on the zeros row of the control matrix.

2.3 NUMERICAL EXAMPLES

In this section, examples are provided to demonstrate the validity of our proposed approach.

Consider the system

$$x(k+1) = \begin{bmatrix} -1.4x_1^2(k) + x_2(k) \\ 0.3x_1(k) + 0 \end{bmatrix} + \begin{bmatrix} 1 \\ 0 \end{bmatrix} u. \tag{2.69}$$

Remark 9. For the system described in (2.69), if we choose the control input, $u(k)$ to be of the form $u(k) = u(k) + 1$, then the system dynamic of (2.69) is equivalent to the Hennon map system. It is well known that for the given parameters (described in (2.69)), the Hennon map behaves chaotically. Refer to Fig. 2.1 for the open-loop response of system (2.69).

Next, with the incorporation of an integrator into the controller structure, (2.69) can be written as

$$\hat{x}(k+1) = \begin{bmatrix} -1.4x_1 & 1 & 1 \\ 0.3 & 0 & 0 \\ 0 & 0 & 1 \end{bmatrix} \begin{bmatrix} x_1(k) \\ x_2(k) \\ x_c(k) \end{bmatrix} + \begin{bmatrix} 0 \\ 0 \\ 1 \end{bmatrix} A_c(x, x_c). \tag{2.70}$$

Furthermore, we choose $\epsilon_1 = \epsilon_2 = 0.01$, and using the procedure described in Corollary 1 with the degree of $\hat{P}(x(k))$ and $\hat{G}(\hat{x}(k))$ set to be 4 and $\hat{L}(\hat{x}(k))$ chosen to be of degree 8, we obtain a feasible solution. The $P(x(k))$ in this work is defined as a symmetrical $N \times N$ polynomial matrix whose

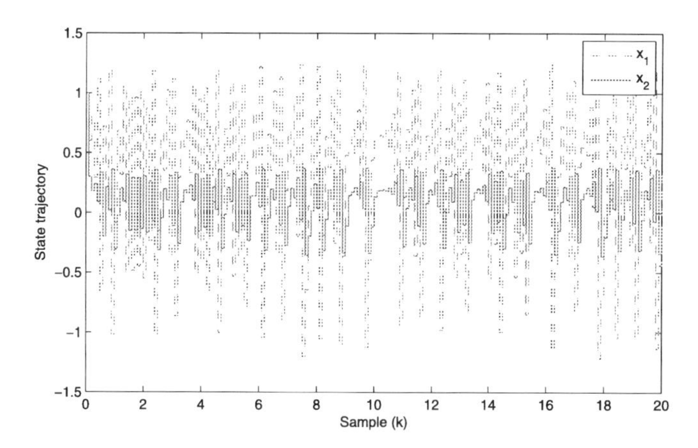

FIGURE 2.1 The trajectory for the open-loop uncertain Hennon Map with $a = 1.4$ and $b = 0.3$.

(i, j)th entry is given by

$$p_{ij}(x(k)) = p_{ij}^{0} + p_{ijg}m(k)^{(1:l)}, \qquad (2.71)$$

where $i = 1, 2, \ldots, n$, $j = 1, 2, \ldots, n$, and $g = 1, 2, \ldots, d$, n is the number of states, d is the total monomial numbers, and $m(k)$ is all monomials vector in $(x(k))$ from degree of 1 to degree of l, where l is a scalar of even value. For example, if $l = 2$ and $x(k) = [x_1(k), x_2(k)]^T$, then $p_{11} = p_{11} + p_{112}x_1 + p_{113}x_2 + p_{114}x_1^2 + p_{115}x_1x_2 + p_{116}x_2^2$. This representation is more general compared to [4] because of a higher value of l, a greater relaxation in the SOS problem can be achieved. The simulation result has been plotted in Fig. 2.2 for the initial value of $[x_1, x_2] = [1, 1]$. From Fig. 2.2, the controller stabilizes the system states to the desired operating region. The controller output response is given in Fig. 2.3.

Remark 10. It is important to highlight here that using the approach proposed in [4], no solution could be obtained for this example. This confirms that our approach is less conservative than [4].

Example 1. **Parametric Uncertainties.** The dynamics of one polynomial discrete-time system is described as follows:

$$x(k+1) = \begin{bmatrix} -ax_1^2(k) + x_2(k) \\ bx_1(k) + 0 \end{bmatrix} + \begin{bmatrix} 1 \\ 0 \end{bmatrix} u(k), \qquad (2.72)$$

where $a = 1.4$, $b = 0.3$. $x_1(k)$ and $x_2(k)$ are the state variables, and $u(k)$ is the control input associated with the system. Next, we assume that there is a $\pm 10\%$ change from their nominal values in its parameter. Therefore, (2.72)

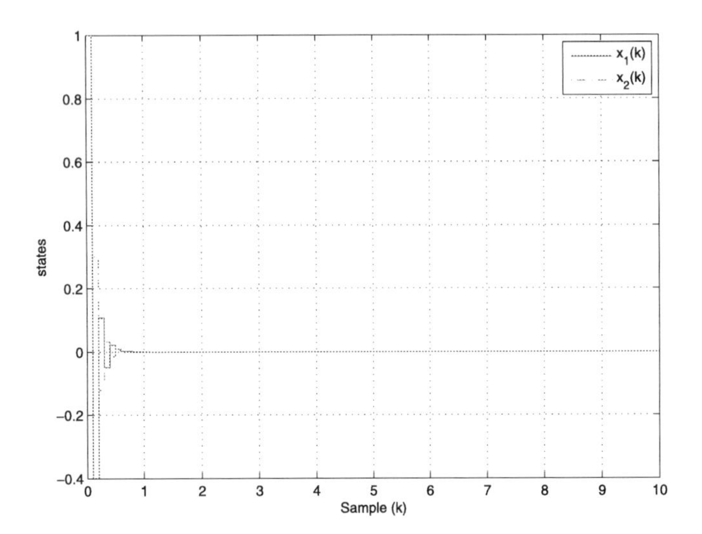

■ **FIGURE 2.2** Response of plant states for nonlinear feedback controller.

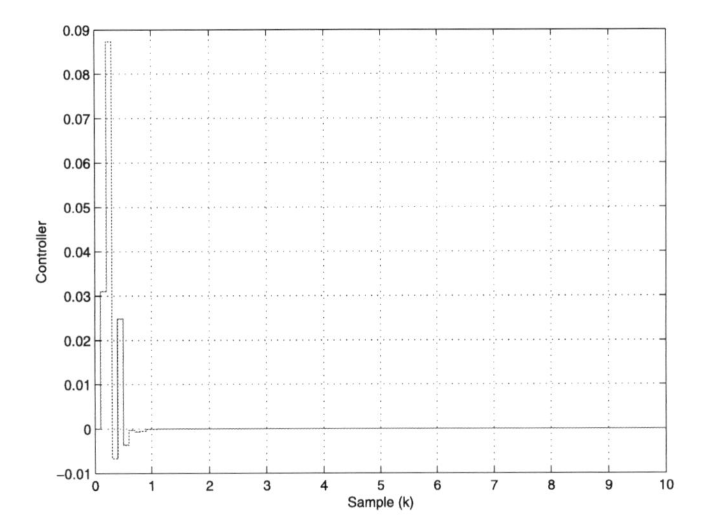

■ **FIGURE 2.3** Controller responses for nonlinear feedback control.

can be written in the form of upper bounds $x_U(k+1)$ and lower bounds $x_L(k+1)$:

$$x_U(k+1) = \begin{bmatrix} -1.26x_1(k) & 1 \\ 0.33 & 0 \end{bmatrix} \begin{bmatrix} x_1(k) \\ x_2(k) \end{bmatrix} + \begin{bmatrix} 1 \\ 0 \end{bmatrix} u(k),$$

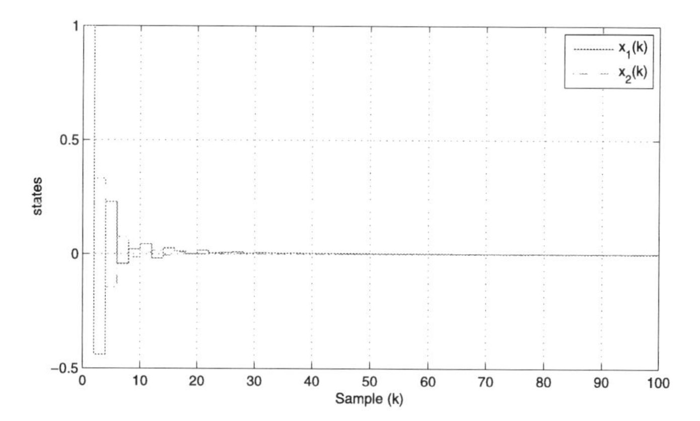

FIGURE 2.4 Response of the plant states for the uncertain system.

$$x_L(k+1) = \begin{bmatrix} -1.54x_1(k) & 1 \\ 0.27 & 0 \end{bmatrix} \begin{bmatrix} x_1(k) \\ x_2(k) \end{bmatrix} + \begin{bmatrix} 1.1 \\ 0 \end{bmatrix} u(k). \tag{2.73}$$

Then, with the integrator in the controller structure, Eq. (2.73) can be rewritten as follows:

$$\begin{aligned} \hat{x}_U(k+1) &= \begin{bmatrix} -1.26x_1(k) & 1 & 1 \\ 0.33 & 0 & 0 \\ 0 & 0 & 1 \end{bmatrix} \begin{bmatrix} x_1(k) \\ x_2(k) \\ x_c(k) \end{bmatrix} + \begin{bmatrix} 0 \\ 0 \\ 1 \end{bmatrix} A_c(x, x_c), \\ \hat{x}_L(k+1) &= \begin{bmatrix} -1.54x_1(k) & 1 & 1.1 \\ 0.27 & 0 & 0 \\ 0 & 0 & 1 \end{bmatrix} \begin{bmatrix} x_1(k) \\ x_2(k) \\ x_c(k) \end{bmatrix} + \begin{bmatrix} 0 \\ 0 \\ 1 \end{bmatrix} A_c(x, x_c). \end{aligned} \tag{2.74}$$

Next, we select $\epsilon_1 = \epsilon_2 = 0.01$, and implementing the procedure outlined in Theorem 3, we initially choose the degree of $\hat{P}_1(x(k))$ and $\hat{P}_2(x(k))$ to be 2. The common polynomial matrices $\hat{G}(\hat{x}(k))$ and $\hat{L}(\hat{x}(k))$ are also selected to be of degree 2, but no feasible solution can be achieved. Then, the degree of $\hat{P}_1(x(k))$, $\hat{P}_2(x(k))$, and $\hat{G}(\hat{x}(k))$ are increased to 4. Meanwhile, the degree of $\hat{L}(\hat{x}(k))$ is increased to 8. With this arrangement, a feasible solution is obtained. The simulation results for this example are given in Fig. 2.4 and Fig. 2.5 between two vertices of (2.74) with the initial condition of $[1 \quad 1]$. It is obvious from Fig. 2.5 that our proposed controller well stabilizes the system.

Remark 11. In [4], the results are produced without consideration of the uncertainties in the system structure. In contrast, we provide the solution for robust control problem with the existence of parametric uncertainties.

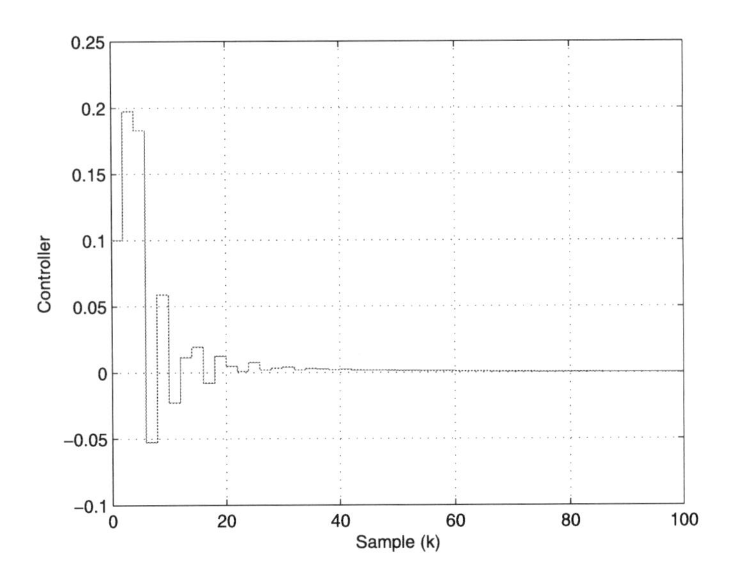

■ **FIGURE 2.5** Controller responses for the uncertain system.

Remark 12. It is confirmed by the simulation examples that by incorporating the integrator into the controller structure a less conservative design procedure can be rendered. However, the price we pay is the large computational cost, which prevents us from using this method for a high-order systems due to the lack of memory space in our machine. This is the main source of conservatism of this integrator method because the size of system matrices will increase. For example, if the original size of one system matrix is 2×2, using this integrator method, the matrix dimension will increase to 3×3. Therefore, if we are dealing with higher-order systems, the computational complexity becomes more severe.

Example 2. **Norm-Bounded Uncertainties.** Consider the following uncertain polynomial discrete-time system:

$$\begin{aligned} x(k+1) = {} & A(x(k))x(k) + \Delta A(x(k))x(k) + B_u(x(k))u(k) \\ & + \Delta B_u(x(k))u(k), \end{aligned} \tag{2.75}$$

where

$$A(x(k)) = \begin{bmatrix} 1 & -T \\ T(1 + a x_1 x_2) & 1 - T \end{bmatrix} \quad \text{and} \quad B_u(x(k)) = \begin{bmatrix} 0 \\ 0.05 \end{bmatrix} \tag{2.76}$$

with $T = 0.05$ and $a = 1$. We assume that the parameters a vary $\pm 30\%$ of their nominal values. Expressing (2.75) in the form (2.46), we have

$$A(x(k)) = \begin{bmatrix} 1 & -T \\ T(1+ax_1x_2) & 1-T \end{bmatrix}; \quad B_u(x(k)) = \begin{bmatrix} 0 \\ 0.05 \end{bmatrix};$$
$$H(x(k)) = \begin{bmatrix} 0 \\ 0.3 \end{bmatrix}; \quad E_1(x(k)) = \begin{bmatrix} ax_1x_2 & 0 \end{bmatrix}; \quad E_2(x(k)) = 0. \quad (2.77)$$

Furthermore, with the incorporation of an integrator into the controller structure, (2.77) can be written as an augmented system,

$$\hat{x}(k+1) = \begin{bmatrix} 1 & -T & 0 \\ T(1+ax_1x_2) & 1-T & 0.05 \\ 0 & 0 & 1 \end{bmatrix} \begin{bmatrix} x_1 \\ x_2 \\ x_c \end{bmatrix}$$
$$+ \begin{bmatrix} 0 \\ 0 \\ 1 \end{bmatrix} A_c(x, x_c) + \Delta\hat{A}(\hat{x}(k)), \quad (2.78)$$

and the uncertainties are described as follows:

$$\Delta\hat{A}(\hat{x}(k)) = \begin{bmatrix} 0 \\ 0.3 \\ 0 \end{bmatrix} \hat{F}(\hat{x}(k)) \begin{bmatrix} ax_1x_2 & 0 & 0 \end{bmatrix}. \quad (2.79)$$

In this example, we choose $\epsilon_1 = \epsilon_2 = 0.01$. The $\epsilon(\hat{x}(k))$ is selected to be of $2\big(x_1^2(k) + x_2^2(k) + x_c^2(k)\big)$. Then, using the procedure described in Proposition 2 with $\hat{P}(x(k))$ and $\hat{G}(\hat{x}(k))$ of degree 4 and $\hat{L}(\hat{x}(k))$ of degree 8, a feasible solution is obtained. The simulation result has been plotted in Fig. 2.6. The initial value for this example is $[x_1, x_2] = [-0.5, 0.5]$. The controller response is shown in Fig. 2.7.

Remark 13. The selection of the design variable $\epsilon(\hat{x}(k))$ is a very crucial in this approach. A wrong selection of this value will drive the solution to be infeasible. In this work, the selection of the value of $\epsilon(\hat{x}(k))$ is delivered using a trial-and-error method.

Remark 14. In this example, we use Simulink to help us to compute $\hat{A}_c(\hat{x}(k))$ from the obtained $\hat{G}(\hat{x}(k))$ and $\hat{L}(\hat{x}(k))$. Due to the large polynomial matrices that resulted from $\hat{A}_c(\hat{x}(k))$, $\hat{G}(\hat{x}(k))$, and $\hat{L}(\hat{x}(k))$, those values are omitted here.

Remark 15. The computational complexity of our proposed method increases with the increment of system's order. This is because the augmented system is used in the problem formulation rather than original system matrices. This is the drawback of this approach because it requires a very large

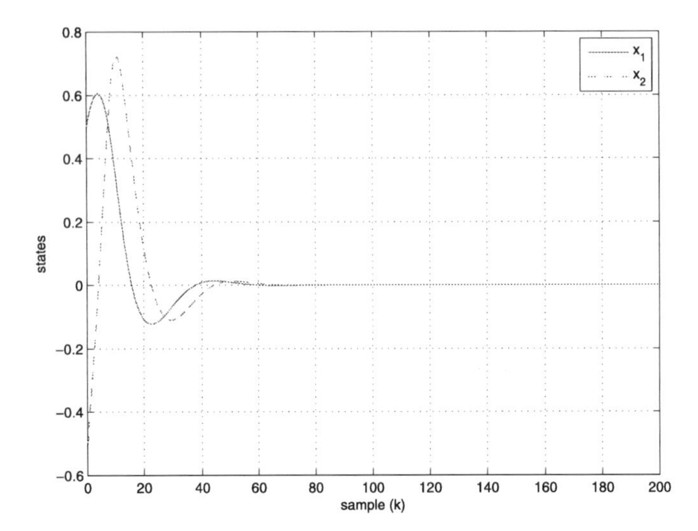

■ **FIGURE 2.6** System states for uncertain discrete-time systems.

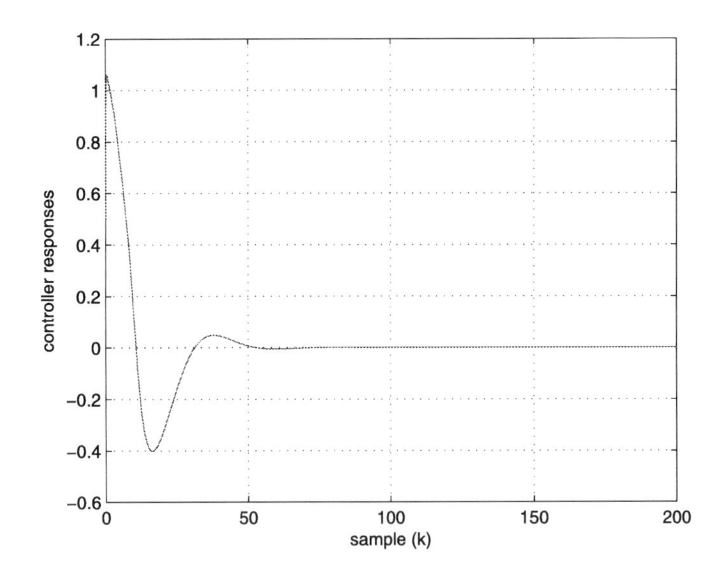

■ **FIGURE 2.7** Controller responses.

of memory spaces to accommodate the computational issue of high-order polynomial systems.

2.4 CONCLUSION

This chapter has examined the problem of designing a nonlinear feedback controller for polynomial discrete-time systems without and with polytopic uncertainties using state-dependent Lyapunov functions. More precisely, we proposed a nonlinear feedback controller with integrator to stabilize such uncertain systems. It has been shown that with the incorporation of the integrator into the controller structure, a less conservative design procedure can be achieved. This is because the nonconvex term $P(x(k+1))$ does not need to be assumed to be bounded and the Lyapunov function does not have to be of a certain form in order to render a convex solution as required in [4]. However, the price we pay is a large computational cost, which prevents us applying this method to the high-order systems. The existence of the proposed nonlinear feedback controller is given in terms of solvability conditions of the polynomial matrix inequalities (PMIs), which are formulated as SOS constraints. The resulting controller gains are in the form of rational matrix function of the augmented states.

REFERENCES

[1] S. Prajna, A. Papachristodoulou, F. Wu, Nonlinear control synthesis by sum of squares optimization: a Lyapunov-based approach, in: Proceedings of the 5th Asian Control Conference, 2004, pp. 157–165.

[2] H.J. Ma, G.H. Yang, Fault-tolerant control synthesis for a class of nonlinear systems: sum of squares optimization approach, International Journal of Robust and Nonlinear Control 19 (5) (2009) 591–610.

[3] D. Zhao, J. Wang, An improved H_∞ synthesis for parameter-dependent polynomial nonlinear systems using SOS programming, in: American Control Conference, 2009, pp. 796–801.

[4] H.J. Ma, G.H. Yang, Fault tolerant H_∞ control for a class of nonlinear discrete-time systems: using sum of squares optimization, in: Proceeding of American Control Conference, 2008, pp. 1588–1593.

[5] J. Doyle, B. Francis, A. Tannenbaum, Feedback Control Theory, Macmillan Publishing, 1990.

[6] G. Dullerud, F. Paganini, A Course in Robust Control Theory, Springer, 1999.

[7] P. Dorato, Robust Control, IEEE Press Book, 1987.

[8] Peng Shi, Shyh-Pyng Shue, Robust H_∞ control for linear discrete-time systems with norm-bounded nonlinear uncertainties, IEEE Transactions on Automatic Control 44 (1) (1999).

[9] J. Zhang, M. Fei, Analysis and design of robust fuzzy controllers and robust fuzzy observers of nonlinear systems, in: Proceedings of the 6th World Congress on Intelligent Control and Automation, Dalian, China, 2006, pp. 3767–3771.

[10] D. Huang, S.K. Nguang, Robust H_∞ static output feedback controller design for fuzzy systems: an ILMI approach, IEEE Transaction on Systems, Man and Cybernetics – Part B: Cybernetics 36 (2006) 216–222.

[11] F. Rasool, D. Huang, S.K. Nguang, Robust H_∞ output feedback control of networked control systems with multiple quantizers, Journal of the Franklin Institute 349 (3) (2012) 1153–1173.
[12] S.K. Nguang, W. Assawinchaichote, P. Shi, Y. Shi, Robust H_∞ control design for uncertain fuzzy systems with Markovian jumps: an LMI approach, in: American Control Conference, 2005, pp. 1805–1810.
[13] S.K. Nguang, P. Zhang, S.X. Ding, Parity relation based fault estimation for nonlinear systems: an LMI approach, International Journal of Automation and Computing 4 (2) (2007) 164–168.
[14] S.K. Nguang, Comments on "Robust stabilization of uncertain input-delay systems by sliding mode control with delay compensation", Automatica 37 (10) (2001) 1677.
[15] S.K. Nguang, P. Shi, H_∞ output feedback control of fuzzy system models under sampled measurements, Computers and Mathematics With Applications 46 (5) (2003) 705–717.
[16] W. Assawinchaichote, S.K. Nguang, P. Shi, Fuzzy Control and Filter Design for Uncertain Fuzzy Systems, Springer, 2006.
[17] S.K. Nguang, P. Shi, On designing filters for uncertain sampled-data nonlinear systems, Systems & Control Letters 41 (5) (2000) 305–316.
[18] D. Huang, S.K. Nguang, Robust Control for Uncertain Networked Control Systems With Random Delays, Springer Science & Business Media, 2009.
[19] J. Zhang, A.K. Swain, S.K. Nguang, Robust sensor fault estimation scheme for satellite attitude control systems, Journal of the Franklin Institute 350 (9) (2013) 2581–2604.
[20] J. Zhang, P. Shi, J. Qiu, S.K. Nguang, A novel observer-based output feedback controller design for discrete-time fuzzy systems, IEEE Transactions on Fuzzy Systems 23 (1) (2015) 223–229.
[21] S.K. Nguang, P. Shi, Delay-dependent H_∞ filtering for uncertain time delay nonlinear systems: an LMI approach, IET Control Theory & Applications 1 (1) (2007) 133–140.
[22] Z. Hou, J. Luo, P. Shi, S.K. Nguang, Stochastic stability of Ito differential equations with semi-Markovian jump parameters, IEEE Transactions on Automatic Control 51 (8) (2006) 1383–1387.
[23] W. Assawinchaichote, S.K. Nguang, P. Shi, E.K. Boukas, H_∞ fuzzy state-feedback control design for nonlinear systems with stability constraints: an LMI approach, Mathematics and Computers in Simulation 78 (4) (2008) 514–531.
[24] S. Chae, S.K. Nguang, SOS based robust H_∞ fuzzy dynamic output feedback control of nonlinear networked control systems, IEEE Transactions on Cybernetics 44 (7) (2014) 1204–1213.
[25] F. Rasool, D. Huang, S.K. Nguang, Robust H_∞ output feedback control of discrete-time networked systems with limited information, Systems & Control Letters 60 (10) (2011) 845–853.
[26] S.K. Nguang, P. Shi, Stabilisation of a class of nonlinear time-delay systems using fuzzy models, in: Proceedings of the 39th IEEE Conference on Decision and Control, 2000, pp. 5–11.
[27] S. Saat, S.K. Nguang, Nonlinear H_∞ output feedback control with integrator for polynomial discrete-time systems, International Journal of Robust and Nonlinear Control 25 (2015) 1051–1065.

[28] Y. Zhang, P. Shi, S.K. Nguang, H.R. Karimi, Observer-based finite-time fuzzy H_∞ control for discrete-time systems with stochastic jumps and time-delays, Signal Processing 97 (2014) 252–261.

[29] S. Chae, F. Rasool, S.K. Nguang, A. Swain, Robust mode delay-dependent H_∞ control of discrete-time systems with random communication delays, IET Control Theory & Applications 4 (6) (2010) 936–944.

[30] Y. Zhang, P. Shi, S.K. Nguang, Y. Song, Robust finite-time H_∞ control for uncertain discrete-time singular systems with Markovian jumps, IET Control Theory & Applications 8 (12) (2014) 1105–1111.

[31] Yong-Yan Cao, P.M. Frank, Robust H_∞ disturbance attenuation for a class of uncertain discrete-time fuzzy systems, IEEE Transactions on Fuzzy Systems 8 (4) (2000).

[32] Peng Shi, Ramesh K. Agarwal, E.K. Boukas, Shyh-Pyng Shue, Robust H_∞ state feedback control of discrete time-delay linear systems with norm-bounded uncertainty, International Journal of System Science 31 (4) (2000) 409–415.

[33] S. Prajna, A. Papachristodoulou, P.A. Parrilo, Introducing SOSTOOLS: a general purpose sum of squares programming solver, in: Conference on Decision and Control, vol. 1, 2002, pp. 741–746.

[34] M.C. de Oliveira, J. Bernussou, J.C. Geromel, A new discrete-time robust stability condition, Systems & Control Letters 37 (4) (1999) 261–265.

[35] P.A. Parrilo, Structured Semidefinite Programs and Semialgebraic Geometry Methods in Robustness and Optimization, PhD dissertation, California Inst. Technol., Pasadena, 2000.

[36] J. Xu, L. Xie, Y. Wang, Synthesis of discrete-time nonlinear systems: a SOS approach, in: American Control Conference ACC '07, 2007, pp. 4829–4834.

[37] B.R. Barmish, Stabilization of uncertain systems via linear control, IEEE Transactions on Automatic Control 28 (8) (1983).

[38] K. Zhou, P.P. Khargonekar, An algebraic Riccati equation approach to H_∞ optimization, System & Control Letters 11 (2) (1988) 85–91.

[39] H.K. Khalil, Nonlinear Systems, Prentice-Hall, 1996.

[40] A. Papachristodoulou, S. Prajna, A tutorial on sum of squares techniques for systems analysis, in: American Control Conference, vol. 4, 2005, pp. 2686–2700.

Chapter 3

Robust nonlinear H_∞ state feedback control for polynomial discrete-time systems

CHAPTER OUTLINE

3.1 INTRODUCTION

The problem of designing a nonlinear H_∞ controller has attracted considerable attention for more than three decades; see, for instance, [1–27]. Generally speaking, the aim of an H_∞ control problem is to design a controller such that the resulting closed-loop control system is stable and a prescribed level of attenuation from the exogenous disturbance input to the output in L_2/l_2-norm is fulfilled. There are two common approaches available to address nonlinear H_∞ control problems: One approach is based on the dissipativity theory [5] and theory of differential games [1]. The other is based on the nonlinear version of the classical bounded real lemma as developed in [6] and [7]. The underlying idea behind both approaches is the conversion of the nonlinear H_∞ control problem into the solvability form of the so-called Hamilton–Jacobi equation (HJE). Unfortunately, this repre-

Analysis and Synthesis of Polynomial Discrete-Time Systems. DOI: 10.1016/B978-0-08-101901-6.00003-7

sentation is hard to solve, and it is generally very difficult to find a global solution.

However, when a polynomial system is under consideration, there is an approach to relax the above-mentioned problem. The approach is called the sum-of-squares (SOS) decomposition and has been developed in [28]. Based on this SOS method, several results can be found in the framework of H_∞ control of polynomial continuous-time systems [29–33]. However, unlike its continuous-time systems counterpart, there are only a few results available that consider the polynomial discrete-time system with H_∞ performance objectives; see [34]. Unfortunately, the aforementioned results suffer from their own conservatism, and such conservatism has been discussed in detail in Chapter 2.

In this chapter, by utilizing the integrator approach as proposed in the previous chapters, we attempt to propose a less conservative design procedure than the available approaches and consequently provide a more general result in the framework of H_∞ control of polynomial discrete-time systems. The result is subsequently extended to a robust H_∞ control problem with the existence of the polytopic uncertainties. The attention here is to design a nonlinear feedback controller such that both stability and a prescribed disturbance attenuation for the closed-loop polynomial discrete-time system are achieved. Furthermore, by an SOS-based method the existence of the proposed controller is given in terms of the solvability of polynomial matrix inequalities (PMIs), which are formulated as SOS constraints and can be solved by the recently developed SOS solvers.

3.2 SYSTEM DESCRIPTION AND PROBLEM FORMULATION

3.2.1 System description

We consider the following dynamic model of a polynomial discrete-time system:

$$\begin{cases} x(k+1) = A(x(k))x(k) + B_u(x(k))u(k) + B_w(x(k))\omega(k), \\ z(k) = C_z(x(k))x(k) + D_{zu}(x(k))u(k), \end{cases} \tag{3.1}$$

where $x(k) \in R^n$ is a state vector, the other vectors are monomials, and $A(x(k))$, $B_u(x(k))$, $B_w(x(k))$, $C_z(x(k))$, and $D_{zu}(x(k))$ are polynomial matrices of appropriate dimensions. In addition, $z(k)$ is a vector of output signals related to the performance of the control system, and $\omega(k)$ is the disturbance belonging to $L_2[0, \infty]$.

We propose the following nonlinear feedback controller with an integrator:

$$\begin{cases} x_c(k+1) = x_c(k) + A_c(x, x_c), \\ u(k) = x_c(k), \end{cases} \tag{3.2}$$

where $A_c(x, x_c)$ is the input function of the integrator, x_c is the controller state, and $u(k)$ is the input function to the system.

Remark 16. The main reason for selecting the controller of the form (3.2) is to ensure that a convex solution to $P(x(k+1))$ can be obtained. A detailed explanation can be found in Chapter 2. Hence we omit the complete discussion here.

System (3.1) with controller (3.2) can now be described as follows:

$$\begin{cases} \hat{x}(k+1) = \hat{A}(\hat{x}(k))\hat{x}(k) + \hat{B}_u(\hat{x}(k))A_c(x, x_c) + \hat{B}_\omega(\hat{x}(k))\omega(k), \\ z(k) = \hat{C}_z(\hat{x}(k))\hat{x}(k), \end{cases} \tag{3.3}$$

where

$$\hat{A}(\hat{x}(k)) = \begin{bmatrix} A(x(k)) & B(x(k)) \\ 0 & 1 \end{bmatrix}; \quad \hat{B}_u(\hat{x}(k)) = \begin{bmatrix} 0 \\ 1 \end{bmatrix};$$

$$\hat{B}_\omega(\hat{x}(k)) = \begin{bmatrix} B_\omega(x(k)) \\ 0 \end{bmatrix}; \quad \hat{C}_z(\hat{x}(k)) = \begin{bmatrix} C_z(x(k))x(k) & D_{zu}(x(k)) \end{bmatrix};$$

$$\hat{x} = \begin{bmatrix} x(k) \\ x_c(k) \end{bmatrix}. \tag{3.4}$$

Next, we assume $A_c(x, x_c)$ to be of the form $A_c(x, x_c) = \hat{A}_c(\hat{x}(k))\hat{x}(k)$. Therefore, (3.3) can be rewritten as follows:

$$\begin{cases} \hat{x}(k+1) = \hat{A}(\hat{x}(k))\hat{x}(k) + \hat{B}_u(\hat{x}(k))\hat{A}_c(\hat{x}(k))\hat{x}(k) + \hat{B}_\omega(\hat{x}(k))\omega(k), \\ z(k) = \hat{C}_z(\hat{x}(k))\hat{x}(k), \end{cases} \tag{3.5}$$

where $\hat{A}(\hat{x}(k))$, $\hat{B}_u(\hat{x}(k))$, $\hat{B}_\omega(\hat{x}(k))$, and $\hat{C}_z(\hat{x}(k))$ are as described in (3.4).

3.2.2 Problem formulation

Problem Formulation: Given a prescribed H_∞ performance $\gamma > 0$, design a nonlinear feedback controller (3.2) such that

$$\|z(k)\|_{[0,\infty]} \leq \gamma^2 \|\omega(k)\|_{[0,\infty]} \tag{3.6}$$

and system in (3.1) with (3.2) is globally asymptotically stable.

3.3 MAIN RESULTS

In this section, we first present a result on the nonlinear H_∞ control problem. Then we subsequently extend it to the robust nonlinear H_∞ control problem with the existence of the uncertainties.

3.3.1 Nonlinear H_∞ control of polynomial discrete systems

Sufficient conditions for the existence of a nonlinear feedback controller of the form (3.2) that satisfies (3.6) are given in the following theorem.

Theorem 5. *Given a prescribed H_∞ performance, $\gamma > 0$, system* (3.1) *is stabilizable with H_∞ performance* (3.6) *via the nonlinear feedback controller of the form* (3.2) *if there exist a symmetric polynomial matrix $\hat{P}(x(k))$ and polynomial matrices $\hat{L}(\hat{x}(k))$ and $\hat{G}(\hat{x}(k))$ such that the following conditions are satisfied for all $x \neq 0$:*

$$\hat{P}(x(k)) > 0, \tag{3.7}$$

$$M(\hat{x}(k)) > 0, \tag{3.8}$$

where

$$M(\hat{x}(k)) = \begin{bmatrix} \hat{G}^T(\hat{x}(k)) + \hat{G}(\hat{x}(k)) - \hat{P}(x(k)) & * & * & * \\ 0 & \gamma^2 I & * & * \\ \hat{A}(\hat{x}(k))\hat{G}(x(k)) + \hat{B}_u(\hat{x}(k))\hat{L}(\hat{x}(k)) & \hat{B}_\omega(\hat{x}(k)) & \hat{P}(x_+) & * \\ \hat{C}_z(\hat{x}(k))\hat{G}(\hat{x}(k)) & 0 & 0 & I \end{bmatrix}. \tag{3.9}$$

The nonlinear feedback controller is given by

$$\begin{aligned} x_c(k+1) =& x_c(k) + A_c(x, x_c), \\ u(k) =& x_c(k), \end{aligned}$$

where $A_c(x, x_c) = \hat{A}_c(\hat{x}(k))\hat{x}(k)$ with $\hat{A}_c(\hat{x}(k)) = \hat{L}(\hat{x}(k))\hat{G}^{-1}(\hat{x}(k))$.

Proof. We select a Lyapunov function of the following form:

$$\hat{V}(\hat{x}(k)) = \hat{x}^T(k)\hat{P}^{-1}(x(k))\hat{x}(k). \tag{3.10}$$

The difference between $\hat{V}(\hat{x}(k+1))$ and $\hat{V}(\hat{x}(k)$ along (3.5) is given by

$$\begin{aligned} \Delta\hat{V}(\hat{x}(k)) =& \hat{x}(k+1)^T \hat{P}^{-1}(x_+)\hat{x}(k+1) - \hat{x}^T(k)\hat{P}^{-1}(x(k))\hat{x}(k) \\ =& \big(\hat{A}(\hat{x}(k))\hat{x}(k) + \hat{B}_u(\hat{x}(k))\hat{A}_c(\hat{x}(k))\hat{x}(k) \end{aligned}$$

$$
\begin{aligned}
&+ \hat{B}_\omega(\hat{x}(k))\omega(k)\big)^T \hat{P}^{-1}(x_+)\big(\hat{A}(\hat{x}(k))\hat{x}(k) \\
&+ \hat{B}_u(\hat{x}(k))\hat{A}_c(\hat{x}(k))\hat{x}(k) + \hat{B}_\omega(\hat{x}(k))\omega(k)\big) \\
&- \hat{x}^T(k)\hat{P}^{-1}(x(k))\hat{x}(k).
\end{aligned} \tag{3.11}
$$

Then, adding and subtracting $-z^T(k)z(k)+\gamma^2\omega^T(k)\omega(k)$ to and from (3.11) result in

$$
\begin{aligned}
&\Delta\hat{V}(\hat{x}(k)) = \big(\hat{A}(\hat{x}(k))\hat{x}(k) + \hat{B}_u(\hat{x}(k))\hat{A}_c(\hat{x}(k))\hat{x}(k) \\
&+ B_\omega(\hat{x}(k))\omega(k)\big)^T \hat{P}^{-1}(x_+)\big(\hat{A}(\hat{x}(k))\hat{x}(k) + \hat{B}_u(\hat{x}(k))\hat{A}_c(\hat{x}(k))\hat{x}(k) \\
&+ B_\omega(\hat{x}(k))\omega((k))\big) - x^T(k)\hat{P}^{-1}(x(k))x(k)) - z^T(k)z(k) + \gamma^2\omega^T(k)\omega(k) \\
&+ z^T(k)z(k) - \gamma^2\omega^T(k)\omega(k) \\
&= \big(\hat{A}(\hat{x}(k))\hat{x}(k) + \hat{B}_u(\hat{x}(k))\hat{A}_c(\hat{x}(k))\hat{x}(k) + \hat{B}_\omega(\hat{x}(k))\omega((k))\big)^T \hat{P}^{-1}(x_+) \\
&\times \big(\hat{A}(\hat{x}(k))\hat{x}(k) + \hat{B}_u(\hat{x}(k))\hat{A}_c(\hat{x}(k))\hat{x}(k) + \hat{B}_\omega(\hat{x}(k))\omega((k))\big) \\
&- x^T(k)\hat{P}^{-1}(x(k))x(k)) + (\hat{C}_z(\hat{x}(k))\hat{x}(k))^T(\hat{C}_z(\hat{x}(k))\hat{x}(k)) \\
&- \gamma^2\omega^T(k)\omega(k) - z^T(k)z(k) + \gamma^2\omega^T(k)\omega(k).
\end{aligned} \tag{3.12}
$$

Now, we write (3.12) as

$$
\Delta\hat{V}(\hat{x}(k)) = \hat{X}^T(k)\Omega(\hat{x}(k))\hat{X}(k) - z^T(k)z(k) + \gamma^2\omega^T(k)\omega(k), \tag{3.13}
$$

where

$$
\Omega(\hat{x}(k)) = \Theta_1(\hat{x}(k))^T\hat{P}^{-1}(x_+)\Theta_1(\hat{x}(k)) + \Theta_2(\hat{x}(k))^T\Theta_2(\hat{x}(k)) - \Lambda
$$

with

$$
\begin{aligned}
\Theta_1(\hat{x}(k)) &= \begin{bmatrix} \hat{A}(\hat{x}(k)) + \hat{B}_u(\hat{x}(k))\hat{A}_c(\hat{x}(k)) & \hat{B}_\omega(\hat{x}(k)) \end{bmatrix}; \\
\Theta_2(\hat{x}(k)) &= \begin{bmatrix} \hat{C}_z(\hat{x}(k)) & 0 \end{bmatrix}; \quad \Lambda = \begin{bmatrix} \hat{P}^{-1}(x(k)) & 0 \\ 0 & \gamma^2 \end{bmatrix}; \\
\hat{X}(k) &= \begin{bmatrix} \hat{x}(k) & \omega(k) \end{bmatrix}^T.
\end{aligned}
$$

Now, we need to show that $\hat{X}^T(k)\Omega(\hat{x}(k))\hat{X}(k) < 0$. To show this, suppose (3.8) is feasible. Then, from the block (1, 1) of (3.9) we have $\hat{G}^T(\hat{x}(k)) + \hat{G}(\hat{x}(k)) > \hat{P}(x(k)) > 0$. This implies that $\hat{G}(\hat{x}(k))$ is nonsingular, and since $\hat{P}(x(k))$ is positive definite, we have

$$
\big(\hat{P}(x(k)) - \hat{G}(\hat{x}(k))\big)^T \hat{P}^{-1}(x_+)\big(\hat{P}(x(k)) - \hat{G}(\hat{x}(k))\big) > 0, \tag{3.14}
$$

and thus

$$\hat{G}^T(\hat{x}(k))\hat{P}^{-1}(x(k))\hat{G}(\hat{x}(k)) \geq \hat{G}^T(\hat{x}(k)) + \hat{G}(\hat{x}(k)) - \hat{P}(x(k)). \quad (3.15)$$

This immediately gives

$$\begin{bmatrix} \hat{G}^T(\hat{x}(k))\hat{P}^{-1}(x(k))\hat{G}(\hat{x}(k)) & * & * & * \\ 0 & \gamma^2 I & * & * \\ \hat{A}(\hat{x}(k))\hat{G}(x(k)) + \hat{B}_u(\hat{x}(k))\hat{L}(\hat{x}(k)) & \hat{B}_\omega(\hat{x}(k)) & \hat{P}(x_+) & * \\ \hat{C}_z(\hat{x}(k))\hat{G}(\hat{x}(k)) & 0 & 0 & I \end{bmatrix} > 0. \quad (3.16)$$

Then, multiplying (3.16) on the right by $diag[\hat{G}^{-1}(\hat{x}(k)), I, I, I]$ and on the left by $diag[\hat{G}^{-1}(\hat{x}(k)), I, I, I]^T$, since $\hat{L}(\hat{x}(k)) = \hat{A}_c(\hat{x}(k))\hat{G}(\hat{x}(k))$, we have

$$\begin{bmatrix} \hat{P}^{-1}(x(k)) & * & * & * \\ 0 & \gamma^2 I & * & * \\ \hat{A}(\hat{x}(k)) + \hat{B}_u(\hat{x}(k))\hat{A}_c(\hat{x}(k)) & \hat{B}_\omega & \hat{P}(x_+) & * \\ \hat{C}_z(\hat{x}(k)) & 0 & 0 & I \end{bmatrix} > 0, \quad (3.17)$$

which is equivalent to

$$\begin{bmatrix} \Lambda & * & * \\ \Theta_1(\hat{x}(k)) & \hat{P}(x_+) & * \\ \Theta_2(\hat{x}(k)) & 0 & I \end{bmatrix} > 0. \quad (3.18)$$

Next, applying the Schur complement to (3.18) results in

$$\left[\Theta_1(\hat{x}(k))^T\hat{P}^{-1}(x_+)\Theta_1(\hat{x}(k)) + \Theta_2(\hat{x}(k))^T\Theta_2(\hat{x}(k)) - \Lambda\right] < 0. \quad (3.19)$$

Then, knowing that (3.19) holds, from (3.13) we have

$$\Delta\hat{V}(\hat{x}(k)) < -z^T(k)z(k) + \gamma^2\omega^T(k)\omega(k). \quad (3.20)$$

Then, the summation from 0 to ∞ yields

$$\hat{V}(\hat{x}(\infty)) - \hat{V}(\hat{x}(0)) \leq -\sum_{k=0}^{\infty} z^T(k)z(k) + \sum_{k=0}^{\infty}\gamma^2\omega^T(k)\omega(k). \quad (3.21)$$

Since $\hat{V}(x(0)) = 0$ and $\hat{V}(x(\infty)) \geq 0$, we obtain

$$\sum_{k=0}^{\infty} z^T(k)z(k) \leq \gamma^2\sum_{k=0}^{\infty}\omega^T(k)\omega(k). \quad (3.22)$$

Hence, (3.6) holds, and therefore the H_∞ performance is fulfilled.

Now we prove that system (3.1) with (3.2) is asymptotically stable. To prove the stability, we set the disturbance $\omega(x(k)) = 0$. Therefore, system (3.1) with controller (3.2) can be described as

$$\hat{x}(k+1) = \hat{A}(\hat{x}(k))\hat{x}(k) + \hat{B}_u(\hat{x}(k))\hat{A}_c(\hat{x}(k))\hat{x}(k), \tag{3.23}$$

where $\hat{A}(\hat{x}(k))$, and $\hat{B}_u(\hat{x}(k))$ are as described in (3.4). From the Lyapunov function described in (3.10), the difference of $\hat{V}(\hat{x}(k+1))$ and $\hat{V}(\hat{x}(k))$ of (3.10) along (3.23) is given by

$$\begin{aligned}
&\Delta\hat{V}(\hat{x}(k)) = \hat{V}(\hat{x}(k+1)) - \hat{V}(\hat{x}(k)) \\
&= \left(\hat{A}(\hat{x}(k))\hat{x}(k) + \hat{B}(\hat{x}(k))\hat{A}_c(\hat{x}(k))\hat{x}(k)\right)^T \hat{P}^{-1}(x_+) \\
&\times \left(\hat{A}(\hat{x}(k))\hat{x}(k) + \hat{B}(\hat{x}(k))\hat{A}_c(\hat{x}(k))\hat{x}(k)\right) - \hat{x}^T(k)\hat{P}^{-1}(x(k))\hat{x}(k) \\
&= \hat{x}^T(k)\big[(\hat{A}^T(\hat{x}(k)) + \hat{A}_c^T(\hat{x}(k))\hat{B}^T(\hat{x}(k)))\hat{P}^{-1}(x_+)(\hat{A}(\hat{x}(k)) \\
&+ \hat{B}(\hat{x}(k))\hat{A}_c(\hat{x}(k))) - \hat{P}^{-1}(x(k))\big]x(k).
\end{aligned} \tag{3.24}$$

By the Lyapunov stability theory [35] system (3.23) is stable if the Lyapunov function (3.10) > 0 is such that (3.24) < 0. Hence, it is obvious from (3.24) that a sufficient condition to achieve (3.24) < 0 is having the terms in $[\cdot] < 0$. Therefore, if

$$\left(\hat{A}^T(\hat{x}) + \hat{A}_c^T(\hat{x})\hat{B}^T(\hat{x})\right)\hat{P}^{-1}(x_+)\left(\hat{A}(\hat{x}) + \hat{B}(\hat{x})\hat{A}_c(\hat{x})\right) - \hat{P}^{-1}(x) < 0, \tag{3.25}$$

then system (3.1) with (3.2) is asymptotically stable. The asymptotic stability of system (3.1) with (3.2) has been proven in Chapter 2. Hence the complete proof is omitted here. □

Note that conditions (3.7)–(3.8) of Theorem 5 are in state-dependent polynomial matrix inequalities (PMIs). Using the SOS decomposition method based on SDP [31] provides a relaxation to the problem. In addition, to ensure the positive definiteness of (3.7) and (3.8), it is often necessary to add some SOS constraints in the form of positive definite constant or polynomial, i.e., $\epsilon > 0$ or $\epsilon(x) > 0$. Then inequalities (3.7)–(3.8) can be modified into SOS as follows:

$$v_1^T[\hat{P}(x(k)) - \epsilon_1 I]v_1 \quad \text{is an SOS,} \tag{3.26}$$

$$v_2^T[M_1(\hat{x}(k)) - \epsilon_2 I]v_2 \quad \text{is an SOS,} \tag{3.27}$$

where v_1 and v_2 are vectors of appropriate dimensions, and $M_1(\hat{x}(k))$ is as defined in (3.8). Moreover, ϵ_1 and ϵ_2 are constant and are always positive, i.e., $\epsilon > 0$. Clearly, this provides sufficient conditions for equations (3.7) and (3.8) and helps the SOS conditions (3.26) and (3.27) to be feasible. Hence, Theorem 5 can be written in the form of SOS conditions, and it is given by the following corollary.

Corollary 2. *Given a prescribed H_∞ performance, $\gamma > 0$, system (3.1) is stabilizable with H_∞ performance (3.6) via the nonlinear feedback controller of the form (3.2) if there exist a symmetric polynomial matrix $\hat{P}(x(k))$, polynomial matrices $\hat{L}(\hat{x}(k))$ and $\hat{G}(\hat{x}(k))$, and constants $\epsilon_1 > 0$ and $\epsilon_2 > 0$ such that the following conditions are satisfied for all $x \neq 0$:*

$$v_1^T[\hat{P}(x(k)) - \epsilon_1 I]v_1 \quad \text{is an SOS,} \tag{3.28}$$

$$v_2^T[M_1(\hat{x}(k)) - \epsilon_2 I]v_2 \quad \text{is an SOS,} \tag{3.29}$$

where $M_1(\hat{x}(k))$ is as given in (3.9), and v_1 and v_2 are free vectors of appropriate dimensions. Moreover, the nonlinear feedback controller is given by

$$\begin{aligned} x_c(k+1) =& x_c(k) + A_c(x, x_c), \\ u(k) =& x_c(k), \end{aligned}$$

where $A_c(x, x_c) = \hat{A}_c(\hat{x}(k))\hat{x}(k)$ with $\hat{A}_c(\hat{x}(k)) = \hat{L}(\hat{x}(k))\hat{G}^{-1}(\hat{x}(k))$.

Remark 17. By using the controller of the form (3.2) and the Lyapunov function of the form (3.10), the nonconvexity $P(x(k+1))$ can be removed efficiently. Hence, Corollary 2 can be solved computationally via SDP. This is the major advantage of our proposed method compared to others.

3.3.2 Robust nonlinear H_∞ control of polynomial discrete systems

In this subsection, we consider both the parametric uncertainty and norm-bounded uncertainties.

3.3.2.1 Parametric uncertainty

$$\begin{aligned} x(k+1) =& A(x(k), \theta)x(k) + B_u(x(k), \theta)u(k) + B_\omega(x(k), \theta)\omega(k), \\ z(k) =& C_z(x(k), \theta)x(k) + D_{zu}(x(k), \theta)u(k), \end{aligned} \tag{3.30}$$

where the matrices $\cdot(x(k), \theta)$ are defined as follows:

$$\begin{aligned}
A(x(k), \theta) &= \sum_{i=1}^{q} A_i(x(k))\theta_i; \quad B(x(k), \theta) = \sum_{i=1}^{q} B_i(x(k))\theta_i; \\
B_\omega(x(k), \theta) &= \sum_{i=1}^{q} B_{\omega i}(x(k))\theta_i; \quad C_z(x(k), \theta) = \sum_{i=1}^{q} C_{zi}(x(k))\theta_i; \\
D_{zu}(x(k), \theta) &= \sum_{i=1}^{q} D_{zui}(x(k))\theta_i.
\end{aligned} \tag{3.31}$$

$\theta = \left[\theta_1, \ldots, \theta_q\right]^T \in \mathbb{R}^q$ is the vector of constant uncertainty and satisfies

$$\theta \in \Theta \triangleq \left\{\theta \in \mathbb{R}^q : \theta_i \geq 0, i = 1, \ldots, q, \sum_{i=1}^{q} \theta_i = 1\right\}. \tag{3.32}$$

We define the nonlinear feedback controller as follows:

$$\begin{cases} x_c(k+1) = x_c(k) + A_c(x, x_c), \\ u(k) = x_c(k), \end{cases} \tag{3.33}$$

where $x_c(k)$ is the state of the controller, and $A_c(x, x_c)$ is the input function yet to be designed.

Formulation of Robust Nonlinear H_∞ Control Problem: Given any $\gamma > 0$, find a controller of the form (3.33) such that the L_2 gain from the disturbance $\omega(k)$ to the output $z(k)$ that needs to be controlled for system (3.30) with (3.33) is less than or equal to γ, i.e.,

$$\|z(k)\|_{[0,\infty]} \leq \gamma^2 \|\omega(k)\|_{[0,\infty]} \tag{3.34}$$

for all $w(k) \in L_2[0, \infty]$ and all admissible uncertainties. In this situation, system (3.41) is said to have a robust H_∞ performance (3.34).

With controller (3.33), we have the following system:

$$\begin{aligned}
\hat{x}(k+1) &= \hat{A}(\hat{x}(k), \theta)\hat{x}(k) + \hat{B}_u(\hat{x}(k), \theta)u(k) + \hat{B}_\omega(\hat{x}(k), \theta)\omega(k), \\
z(k) &= \hat{C}_z(x(k), \theta)\hat{x}(k),
\end{aligned} \tag{3.35}$$

where

$$\hat{A}(\hat{x}(k),\theta)=\sum_{i=1}^{q}\hat{A}_i(\hat{x}(k))\theta_i;\quad \hat{B}_u(\hat{x}(k),\theta)=\sum_{i=1}^{q}\hat{B}_i(\hat{x}(k))\theta_i;$$
$$\hat{C}_z(x(k),\theta)=\sum_{i=1}^{q}\hat{C}_{zi}(x(k))\theta_i;\quad \hat{B}_\omega(\hat{x}(k),\theta)=\sum_{i=1}^{q}\hat{B}_i(\hat{x}(k))\theta_i. \tag{3.36}$$

We further define the following parameter-dependent Lyapunov function:

$$\hat{V}(\hat{x}(k))=\hat{x}^T(k)\Big(\sum_{i=1}^{q}\hat{P}_i(x(k))\theta_i\Big)^{-1}\hat{x}(k), \tag{3.37}$$

where $\hat{P}(x(k))$ is as defined in the previous section.

By the results given in the previous section we have the following theorem.

Theorem 6. *Given a prescribed H_∞ performance $\gamma>0$, constants $\epsilon_1>0$ and $\epsilon_2>0$ for $x\neq 0$ and $i=1,\ldots,q$, system (3.30) with the nonlinear feedback controller (3.2) is asymptotically stable with H_∞ performance (3.6) for $x\neq 0$ if there exist common polynomial matrices $\hat{G}(\hat{x}(k))$ and $\hat{L}(\hat{x}(k))$ and a symmetric polynomial matrix $\hat{P}_i(x(k))$ such that the following conditions are satisfied for all $x\neq 0$:*

$$v_3^T[\hat{P}_i(x(k))-\epsilon_1 I]v_3 \quad \textit{is an SOS}, \tag{3.38}$$
$$v_4^T\left[M_2(\hat{x}(k))-\epsilon_2 I\right]v_4 \quad \textit{is an SOS}, \tag{3.39}$$

where v_3 and v_4 are free vectors of appropriate dimensions.

$$M_2(\hat{x}(k))=\begin{bmatrix}\hat{G}^T(\hat{x}(k))+\hat{G}(\hat{x}(k))-\hat{P}_i(x(k)) & * & * & *\\ 0 & \gamma^2 I & * & *\\ \hat{A}_i(\hat{x}(k))\hat{G}(x(k))+\hat{B}_{ui}(\hat{x}(k))\hat{L}(\hat{x}(k)) & \hat{B}_{\omega i}(\hat{x}(k)) & \hat{P}_i(x_+) & *\\ \hat{C}_{zi}(\hat{x}(k))\hat{G}(\hat{x}(k)) & 0 & 0 & I\end{bmatrix}. \tag{3.40}$$

Moreover, a suitable choice of the controller is given as

$$x_c(k+1)=x_c(k)+A_c(x,x_c),$$
$$u(k)=x_c(k).$$

where $A_c(x,x_c)=\hat{A}_c(\hat{x}(k))\hat{x}(k)$ with $\hat{A}_c(\hat{x}(k))=\hat{L}(\hat{x}(k))\hat{G}^{-1}(\hat{x}(k))$.

Proof. This theorem follows directly as a convex combination of several systems of the form (3.1) for a common (3.33). □

3.3.2.2 Norm-bounded uncertainties

Consider the following polynomial discrete-time system with uncertainties in the state and input:

$$\begin{cases} x(k+1) = A(x(k))x(k) + \Delta A(x(k))x(k) + B_u(x(k))u(k) \\ \qquad\qquad + \Delta B_u(x(k))u(k) + B_\omega(x(k))\omega(k), \\ z(k) = C_z(x(k))x(k) + D_{zu}(x(k))u(k), \end{cases} \tag{3.41}$$

where $x(k) \in R^n$ is the state vector, $u(k) \in R^m$ is the input, $A(x(k))$, $B_u(x(k))$, $C_z(x(k))$, and $D_{zu}(x(k))$ are polynomial matrices of appropriate dimensions, $z(k)$ is a vector of output signals related to the performance of the control system, and $\omega(k)$ is the disturbance belonging to $L_2[0,\infty]$. Meanwhile, $\Delta A(x(k))$ and $\Delta B_u(x(k))$ represent the uncertainties in the system and satisfy the following assumption.

Assumption. The parameter uncertainties considered here are norm-bounded and described by the following form:

$$\begin{bmatrix}\Delta A(x(k)) & \Delta B_u(x(k))\end{bmatrix} = H(x(k))F(x(k))\begin{bmatrix}E_1(x(k)) & E_2(x(k))\end{bmatrix}, \tag{3.42}$$

where $H(x(k))$, $E_1(x(k))$, and $E_2(x(k))$ are known polynomial matrices of appropriate dimensions, and $F(x(k))$ is an unknown matrix function that satisfies

$$\left\| F^T(x(k))F(x(k)) \right\| \leq I. \tag{3.43}$$

Motivated by the work [36], we define the following "*scaled*" system:

$$\begin{cases} \tilde{x}(k+1) = A(\tilde{x}(k))\tilde{x}(k) + \begin{bmatrix} B_\omega(\tilde{x}(k)) & \frac{1}{\delta}\bar{H}(\tilde{x}(k)) \end{bmatrix}\tilde{\omega}(k) \\ \qquad\qquad + B_u(\tilde{x}(k))u(k), \\ \tilde{z}(k) = \begin{bmatrix} C_z(\tilde{x}(k)) \\ \delta E_1(\tilde{x}(k)) \end{bmatrix}\tilde{x}(k) + \begin{bmatrix} D_{zu}(\tilde{x}(k)) \\ \delta E_2(\tilde{x}(k)) \end{bmatrix}u(k), \end{cases} \tag{3.44}$$

where $\tilde{x} \in R^n$ is the state, $u(k) \in R^m$, $\tilde{\omega} \in R^{m+i}$ is the input noise, δ is a positive constant, $\tilde{z}(k)$ is the controlled output, and $\bar{H}(\tilde{x}(k)) = [H_1(\tilde{x}(k)) \quad H_1(\tilde{x}(k))]$.

System (3.44) with controller (3.33) can be written as follows:

$$\begin{cases} \hat{x}(k+1) = \hat{A}(\hat{x}(k))\hat{x}(k) + \hat{B}_u(\hat{x}(k))A_c(x,x_c) + \hat{B}_\omega(\hat{x}(k))\tilde{\omega}(k), \\ \tilde{z}(k) = \hat{C}_z(\hat{x}(k))\hat{x}(k), \end{cases} \tag{3.45}$$

where

$$\hat{A}(\hat{x}(k)) = \begin{bmatrix} A(\tilde{x}(k)) & B_u(\tilde{x}(k)) \\ 0 & 1 \end{bmatrix}, \quad \hat{B}_u(\hat{x}(k)) = \begin{bmatrix} 0 \\ 1 \end{bmatrix},$$

$$\hat{B}_\omega(\hat{x}(k)) = \begin{bmatrix} \tilde{B}_\omega(\tilde{x}(k)) \\ 0 \end{bmatrix}, \quad \hat{C}_z(\hat{x}(k)) = \begin{bmatrix} \tilde{C}_z(\tilde{x}(k)) & \tilde{D}_{zu}(\tilde{x}(k)) \end{bmatrix},$$

$$\hat{x} = \begin{bmatrix} x(k) \\ x_c(k) \end{bmatrix}, \tag{3.46}$$

with

$$\tilde{B}_\omega(\tilde{x}(k)) = \begin{bmatrix} B_\omega(\tilde{x}(k)) & \frac{1}{\delta}\bar{H}(\tilde{x}(k)) \end{bmatrix}, \quad \tilde{C}_z(\tilde{x}(k)) = \begin{bmatrix} C_z(\tilde{x}(k)) \\ \delta E_1(\tilde{x}(k)) \end{bmatrix},$$

$$\tilde{D}_{zu}(\tilde{x}(k)) = \begin{bmatrix} D_{zu}(\tilde{x}(k)) \\ \delta E_2(\tilde{x}(k)) \end{bmatrix}. \tag{3.47}$$

Next, we assume $A_c(x, x_c)$ to be of the form $A_c(x, x_c) = \hat{A}_c(\hat{x}(k))\hat{x}(k)$. Therefore, (3.45) can be rewritten as follows:

$$\begin{cases} \hat{x}(k+1) = \hat{A}(\hat{x}(k))\hat{x}(k) + \hat{B}_u(\hat{x}(k))\hat{A}_c(\hat{x}(k))\hat{x}(k) + \hat{B}_\omega(\hat{x}(k))\tilde{\omega}(k), \\ \tilde{z}(k) = \hat{C}_z(\hat{x}(k))\hat{x}(k), \end{cases} \tag{3.48}$$

where $\hat{A}(\hat{x}(k))$, $\hat{B}_u(\hat{x}(k))$, $\hat{B}_\omega(\hat{x}(k))$, and $\hat{C}_z(\hat{x}(k))$ are as described in (3.46).

In view of the "*scaled*" system (3.44), we establish the following theorem.

Theorem 7. *Consider system* (3.41). *There exists a controller of the form* (3.33) *such that* (3.34) *holds for all admissible uncertainties if there exists a positive constant* $\delta > 0$ *such that* (3.34) *holds for system* (3.44) *with the same controller.*

Proof. Suppose that

$$\|\tilde{z}(k)\|_{[0,\infty]} \le \|\tilde{\omega}(k)\|_{[0,\infty]} \tag{3.49}$$

for (3.44) with (3.33) for all $w(k) \in L_2[0, \infty]$. Then, we need to show that

$$\|z(k)\|_{[0,\infty]} \le \|\omega(k)\|_{[0,\infty]} \tag{3.50}$$

for system (3.41) with the same controller. To show this, let us choose

$$\tilde{\omega} = \begin{bmatrix} \omega(k) \\ \delta\eta(k) \end{bmatrix}, 0 \le k \le \infty, \tag{3.51}$$

where

$$\eta(k) = F(x(k)) \begin{bmatrix} E_1(x(k))x(k) \\ E_2(x(k))u(k) \end{bmatrix}. \tag{3.52}$$

Then, it is trivial to show that (3.49) implies (3.50). This completes the proof. □

In light of Theorem 7, it remains to solve the "*scaled*" nonlinear H_∞ control problem given in (3.44). Therefore, sufficient conditions for the existence of a solution to the robust H_∞ control problem are presented in the following theorem.

Theorem 8. *Given a prescribed H_∞ performance $\gamma > 0$, system (3.41) is stabilizable with H_∞ performance (3.34) via the nonlinear feedback controller of the form (3.33) if there exist a symmetric polynomial matrix $\hat{P}(x(k))$ and polynomial matrices $\hat{L}(\hat{x}(k))$ and $\hat{G}(\hat{x}(k))$ such that the following conditions are satisfied for all $x \neq 0$:*

$$\hat{P}(x(k)) > 0, \tag{3.53}$$

$$M(\hat{x}(k)) > 0, \tag{3.54}$$

where

$$M(\hat{x}(k)) = \begin{bmatrix} \hat{G}^T(\hat{x}(k)) + \hat{G}(\hat{x}(k)) - \hat{P}(x(k)) & * & * & * \\ 0 & \gamma^2 I & * & * \\ \hat{A}(\hat{x}(k))\hat{G}(x(k)) + \hat{B}_u(\hat{x}(k))\hat{L}(\hat{x}(k)) & \hat{B}_\omega(\hat{x}(k)) & \hat{P}(x_+) & * \\ \hat{C}_z(\hat{x}(k))\hat{G}(\hat{x}(k)) & 0 & 0 & I \end{bmatrix}. \tag{3.55}$$

Moreover, a suitable nonlinear feedback controller is given by

$$x_c(k+1) = x_c(k) + A_c(x, x_c),$$
$$u(k) = x_c(k),$$

where $A_c(x, x_c) = \hat{A}_c(\hat{x}(k))\hat{x}(k)$ with $\hat{A}_c(\hat{x}(k)) = \hat{L}(\hat{x}(k))\hat{G}^{-1}(\hat{x}(k))$.

Proof. By Theorem 7 the robust nonlinear H_∞ control problem is converted to the nonlinear H_∞ control problem for a "*scaled*" system. Then, by adapting Theorem 5, the result can be obtained easily. It is important to note here that the Lyapunov function of the following form is selected:

$$\hat{V}(\hat{x}(k)) = \hat{x}^T(k)\hat{P}^{-1}(x(k))\hat{x}(k). \tag{3.56}$$

□

Remark 18. The idea of choosing the Lyapunov function to be of the form (3.56) is to ensure that a convex solution of $P(x(k+1))$ can be achieved. This idea has been outlined in detail in Chapter 2, and hence, for simplicity, the complete explanation is omitted here.

Note that the conditions (3.53)–(3.54) of Theorem 8 are in terms of state-dependent polynomial matrix inequalities (PMIs). Using the SOS decomposition method based on SDP [31] provides a relaxation of the problem.

Corollary 3. *Given a prescribed H_∞ performance $\gamma > 0$, system* (3.41) *is stabilizable with H_∞ performance* (3.34) *via the nonlinear feedback controller of the form* (3.33) *if there exist a symmetric polynomial matrix $\hat{P}(x(k))$, polynomial matrices $\hat{L}(\hat{x}(k))$ and $\hat{G}(\hat{x}(k))$, and constants $\epsilon_1 > 0$ and $\epsilon_2 > 0$ such that the following conditions are satisfied for all $x \neq 0$:*

$$v_1^T[\hat{P}(x(k)) - \epsilon_1 I]v_1 \quad \text{is an SOS,} \tag{3.57}$$

$$v_2^T[M(\hat{x}(k)) - \epsilon_2 I]v_2 \quad \text{is an SOS,} \tag{3.58}$$

where $M(\hat{x}(k))$ is as given in (3.55), *and v_1 and v_2 are free vectors of appropriate dimensions. Moreover, the nonlinear feedback controller is given by*

$$\begin{aligned} x_c(k+1) =& x_c(k) + A_c(x, x_c), \\ u(k) =& x_c(k), \end{aligned}$$

where $A_c(x, x_c) = \hat{A}_c(\hat{x}(k))\hat{x}(k)$ with $\hat{A}_c(\hat{x}(k)) = \hat{L}(\hat{x}(k))\hat{G}^{-1}(\hat{x}(k))$.

3.4 NUMERICAL EXAMPLES

Example 1. Consider a tunnel diode circuit as shown in Fig. 3.1 [37], where the characteristics of the tunnel diode are described as follows:

$$i_D(t) = 0.002v_D(t) + 0.01v_D^3(t). \tag{3.59}$$

Next, letting $x_1(t) = v_c(t)$ and $x_2(t) = i_L(t)$ be the state variables, the circuit is governed by the following state equations:

$$\begin{aligned} C\dot{x}_1(t) &= -0.002x_1(t) - 0.01x_1^3(t) + x_2(t), \\ L\dot{x}_2(t) &= -x_1(t) - Rx_2(t) + \omega(t) + u(t), \\ z(t) &= x_2(t) + u(t), \end{aligned} \tag{3.60}$$

where $\omega(t)$ is the noise to the system, $z(t)$ is the controlled output, and we assume that both $x_1(t) = v_c(t)$ and $x_2(t) = i_L(t)$ are available for

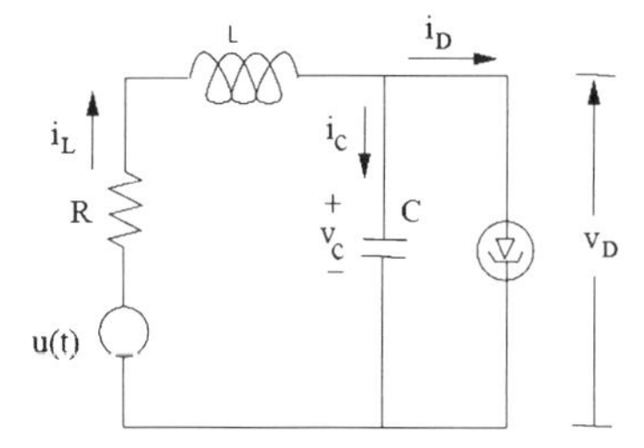

FIGURE 3.1 A tunnel diode circuit.

feedback. Meanwhile, the circuit parameter is given as follows: $C = 20$ mF, $L = 1000$ mH, and $R = 1\,\Omega$. With these parameters, the dynamics of the circuit can be written as follows:

$$\begin{aligned}\dot{x}_1(t) &= -0.1x_1(t) - 0.5x_1^3(t) + 50x_2(t),\\ \dot{x}_2(t) &= -x_1(t) - x_2(t) + \omega(t) + u(t),\\ z(t) &= x_2(t) + u(t).\end{aligned} \tag{3.61}$$

Then, the above system is sampled at $T = 0.02$ and by Euler's discretization method, the following discrete-time nonlinear dynamic equations are obtained:

$$\begin{aligned}x_1(k+1) &= x_1(k) + T\big[-0.1x_1(k) - 0.5x_1^3(k) + 50x_2(k)\big],\\ x_2(k+1) &= x_2(k) + T\big[-x_1(t) - x_2(t) + \omega(t) + u(t)\big],\\ z(k) &= x_2(k) + u(k).\end{aligned} \tag{3.62}$$

From (3.62), the system with controller (3.2) can be written as follows:

$$\begin{cases}\hat{x}(k+1) = \hat{A}(\hat{x}(k))\hat{x}(k) + \hat{B}_u(\hat{x}(k))\hat{A}_c(\hat{x}(k))\hat{x}(k) + \hat{B}_\omega(\hat{x}(k))\omega(k),\\ \quad z(k) = \hat{C}_z(\hat{x}(k))\hat{x}(k),\end{cases} \tag{3.63}$$

where

$$\hat{A}(\hat{x}(k)) = \begin{bmatrix} 1 + T\big(-0.1 - 0.5x_1^2(k)\big) & 50T & 0\\ -T & 1-T & T\\ 0 & 0 & 1\end{bmatrix},$$

$$\hat{B}_u(\hat{x}(k)) = \begin{bmatrix}0\\0\\1\end{bmatrix},\quad \hat{B}_\omega(\hat{x}(k)) = \begin{bmatrix}0\\T\\0\end{bmatrix},\quad \hat{C}_z(\hat{x}(k)) = \begin{bmatrix}0 & 1 & 1\end{bmatrix},$$

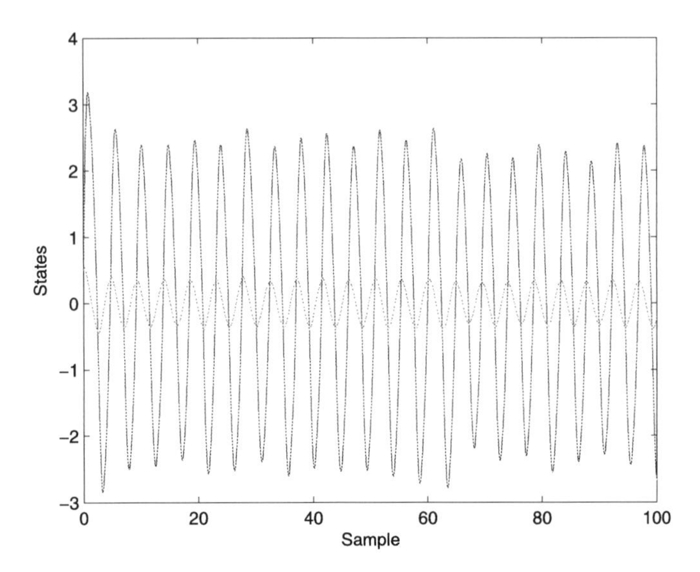

FIGURE 3.2 Open-loop responses for a tunnel diode circuit.

$$\hat{x}(k) = \begin{bmatrix} x_1(k) \\ x_2(k) \\ x_c(k) \end{bmatrix}. \tag{3.64}$$

Remark 19. For this example, we choose $\epsilon_1 = \epsilon_2 = 0.01$ and $\gamma = 1$. The open-loop response is given in Fig. 3.2. Then, using the procedure described in the Corollary 2, with $\hat{P}(x(k))$ and $\hat{G}(\hat{x}(k))$ of degree 4 and $\hat{L}(\hat{x}(k))$ of degree 8, a feasible solution is achieved. The band-limited white noise (noise power = 10) is used in the simulation. The energy ratio of the regulated output and the disturbance input noise is shown in Fig. 3.3. The value shown is less than a prescribed value 1.

Example 2. **Robust nonlinear H_∞ control problem – Parametric Uncertainties.** Consider the tunnel diode circuit shown in Fig. 3.1. For this example, we assume that the value of R is uncertain and given by $R = 1 \pm 30\%\ \Omega$. Therefore, the system can be described as follows:

$$\hat{A}(\hat{x}(k)) = \begin{bmatrix} 1 + T\big(-0.1 - 0.5x_1^2(k)\big) & 50T & 0 \\ -T & 1 - RT & T \\ 0 & 0 & 1 \end{bmatrix},$$

$$\hat{B}_u(\hat{x}(k)) = \begin{bmatrix} 0 \\ 0 \\ 1 \end{bmatrix}, \quad \hat{B}_\omega(\hat{x}(k)) = \begin{bmatrix} 0 \\ T \\ 0 \end{bmatrix}, \quad \hat{C}_z(\hat{x}(k)) = \begin{bmatrix} 0 & 1 & 1 \end{bmatrix}. \tag{3.65}$$

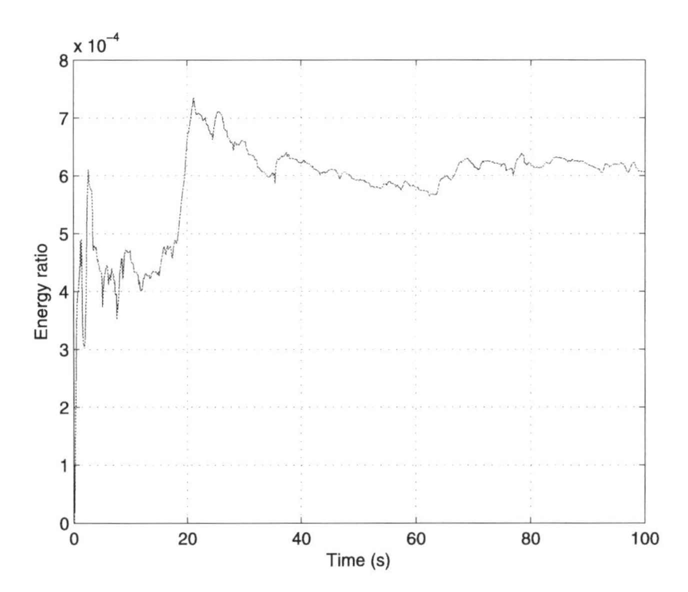

■ **FIGURE 3.3** Ratio of the regulated output energy to the disturbance input noise energy without uncertainty.

By implementing Theorem 6, where γ is chosen to be 1 and the values of the positive constants ϵ_1 and ϵ_2 are fixed at 0.01, we initially choose the degree of $\hat{P}_1(x(k))$ and $\hat{P}_2(x(k))$ to be 2. The common polynomial matrices $\hat{G}(\hat{x}(k))$ and $\hat{L}(\hat{x}(k))$ are also selected to be of degree 2, but no feasible solution can be achieved. Then, the degrees of $\hat{P}_1(x(k))$, $\hat{P}_2(x(k))$, and $\hat{G}(\hat{x}(k))$ are increased to 4. Meanwhile, the degree of $\hat{L}(\hat{x}(k))$ is increased to 8, and with this arrangement, a feasible solution is obtained. From the simulation result shown in Fig. 3.4, the ratio of the regulated output energy to the disturbance input noise energy in this example tends to be a constant value, which is approximately at 1.3×10^{-4}. Thus, $\gamma = \sqrt{1.3 \times 10^{-4}} \approx 0.0114$. This implies that the L_2 gain from the disturbance to the regulated output is less than 0.00114.

Remark 20. It is worth noting here that, to date, no result has been presented in the literature that considers the robust H_∞ stabilization for the polynomial discrete-time system.

Example 3. **Robust nonlinear H_∞ control problem – Norm-Bounded Uncertainties.** Consider a tunnel diode circuit as shown in Fig. 3.1, where the characteristics of the tunnel diode are described as follows:

$$i_D(t) = 0.002v_D(t) + 0.01v_D^3(t). \tag{3.66}$$

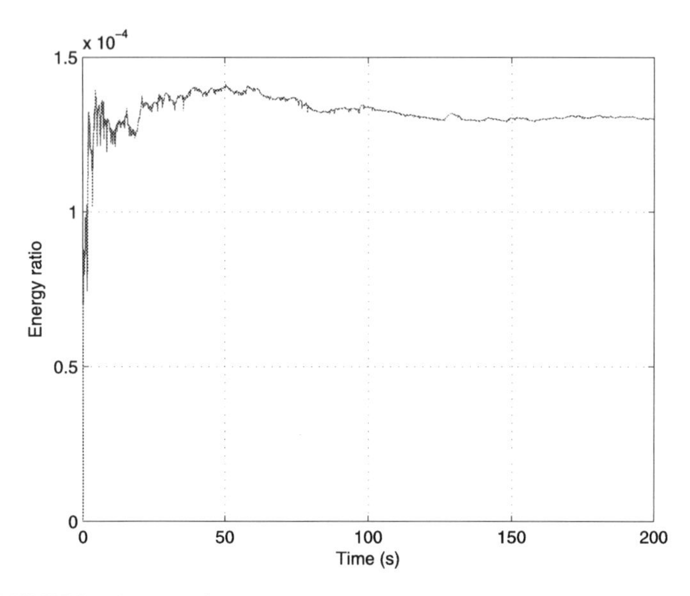

■ **FIGURE 3.4** Energy ratio of the regulated output and the disturbance input noise with polytopic uncertainties.

Meanwhile, the circuit parameter is given as follows: $C = 20$ mF, $L = 1000$ mH, and $R = 1 \pm 30\%\ \Omega$. With these parameters, the dynamics of the circuit can be written as

$$\begin{aligned} \dot{x}_1(t) &= -0.1x_1(t) - 0.5x_1^3(t) + 50x_2(t), \\ \dot{x}_2(t) &= -x_1(t) - (1 + \Delta R)x_2(t) + \omega(t) + u(t), \\ z(t) &= x_2(t) + u(t). \end{aligned} \tag{3.67}$$

Then, the above system is sampled at $T = 0.02$, and by using Euler's discretization method the following discrete-time nonlinear dynamic equations are obtained:

$$\begin{aligned} x_1(k+1) &= x_1(k) + T\big[-0.1x_1(k) - 0.5x_1^3(k) + 50x_2(k)\big], \\ x_2(k+1) &= x_2(k) + T\big[-x_1(k) - (1 + \Delta R)x_2(k) + \omega(k) + u(k)\big], \\ z(k) &= x_2(k) + u(k). \end{aligned} \tag{3.68}$$

From (3.68), the system with controller (3.33) can be written as follows:

$$\begin{cases} \hat{x}(k+1) = \hat{A}(\hat{x}(k))\hat{x}(k) + \hat{B}_u(\hat{x}(k))\hat{A}_c(\hat{x}(k))\hat{x}(k) + \hat{B}_\omega(\hat{x}(k))\omega(k), \\ \quad z(k) = \hat{C}_z(\hat{x}(k))\hat{x}(k), \end{cases} \tag{3.69}$$

where

$$\hat{A}(\hat{x}(k)) = \begin{bmatrix} A(\tilde{x}(k)) & B_u(\tilde{x}(k)) \\ 0 & 1 \end{bmatrix}, \quad \hat{B}_u(\hat{x}(k)) = \begin{bmatrix} 0 \\ 1 \end{bmatrix},$$
$$\hat{B}_\omega(\hat{x}(k)) = \begin{bmatrix} \tilde{B}_\omega(\tilde{x}(k)) \\ 0 \end{bmatrix}, \quad \hat{C}_z(\hat{x}(k)) = \begin{bmatrix} \tilde{C}_z(\tilde{x}(k)) & \tilde{D}_{zu}(\tilde{x}(k)) \end{bmatrix},$$
$$\hat{x} = \begin{bmatrix} x(k) \\ x_c(k) \end{bmatrix}, \tag{3.70}$$

with

$$\tilde{B}_\omega(\tilde{x}(k)) = \begin{bmatrix} B_\omega(\tilde{x}(k)) & \frac{1}{\delta}\bar{H}(\tilde{x}(k)) \end{bmatrix}, \quad \tilde{C}_z(\tilde{x}(k)) = \begin{bmatrix} C_z(\tilde{x}(k)) \\ \delta E_1(\tilde{x}(k)) \end{bmatrix},$$
$$\tilde{D}_{zu}(\tilde{x}(k)) = \begin{bmatrix} D_{zu}(\tilde{x}(k)) \\ \delta E_2(\tilde{x}(k)) \end{bmatrix}. \tag{3.71}$$

Similarly,

$$\hat{A}(\hat{x}(k)) = \begin{bmatrix} 1 + T\left[-0.1 - 0.5x_1^2(k)\right] & 50T & 0 \\ -T & 1 - T(1 + \Delta R) & T \\ 0 & 0 & 1 \end{bmatrix},$$
$$\hat{B}_u(\hat{x}(k)) = \begin{bmatrix} 0 \\ 0 \\ 1 \end{bmatrix}, \quad \hat{B}_\omega(\hat{x}(k)) = \begin{bmatrix} 0 & 0 & 0 \\ T & \frac{1}{\delta}0.3 & \frac{1}{\delta}0.3 \\ 0 & 0 & 0 \end{bmatrix},$$
$$\hat{C}_z(\hat{x}(k)) = \begin{bmatrix} 0 & 1 & 1 \\ 0 & \delta & 0 \end{bmatrix}, \quad \hat{x}(k) = \begin{bmatrix} x_1(k) \\ x_2(k) \\ x_c(k) \end{bmatrix}, \tag{3.72}$$

where $\delta = 1$. From (3.72), for clarity, we describe again the matrices that represent the uncertainty:

$$H(x(k)) = \begin{bmatrix} 0 \\ 0.3 \end{bmatrix}, \quad E_1(x(k)) = \begin{bmatrix} 0 & \delta \end{bmatrix}, \quad E_2(x(k)) = \begin{bmatrix} 0 \end{bmatrix}.$$

Then, applying the procedures outlined in Corollary 3 with $\hat{P}(\hat{x}(k))$ and $\hat{G}(\hat{x}(k))$ of degree 4 and $\hat{L}(\hat{x}(k))$ of degree 8, a feasible solution is obtained. The ratio of the regulated output energy to the noise energy is shown in Fig. 3.5. It can be clearly seen from the figure that the energy ratio tends to be a constant value after 15 s, which is approximately 4.25×10^{-4}. Hence, the value of γ is equivalent to $\sqrt{4.25 \times 10^{-4}} \approx 0.0206$. This value is absolutely less than a prescribed γ value 1.

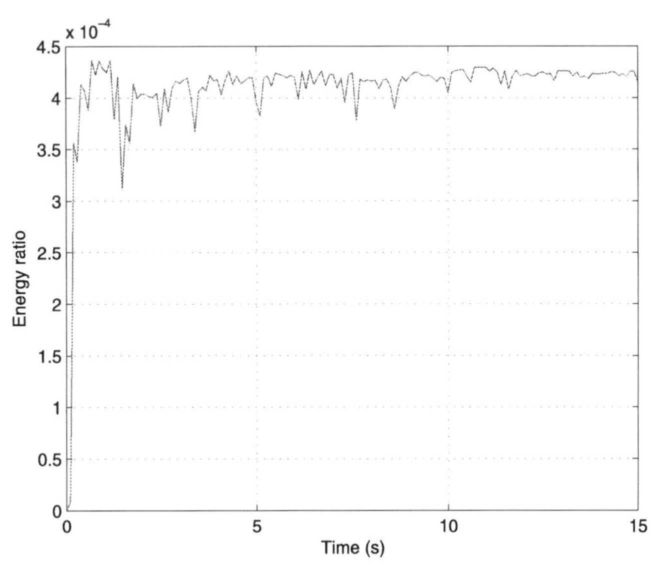

■ **FIGURE 3.5** Energy ratio: $\frac{\sum z^T z}{\sum \omega^T \omega}$.

REFERENCES

[1] J.A. Ball, J.W. Helton, H_∞ control for nonlinear plants: connection with differential games, in: Proc. 28th IEEE Conf. Decision Control, 1989, pp. 956–962.

[2] T. Basar, G.J. Olsder, Dynamic Noncooperative Game Theory, Academic, New York, 1982.

[3] A.J. Van Der Schaft, L_2-gain analysis of nonlinear systems and nonlinear state feedback H_∞ control, IEEE Transactions on Automatic Control 37 (6) (1992) 770–784.

[4] A. Isodori, A. Astolfi, Disturbance attenuation and H_∞ control via measurement feedback in nonlinear systems, IEEE Transactions on Automatic Control 37 (9) (1992) 1283–1293.

[5] T. Basar, Optimum performance levels for minimax filters, predictors and smoothers, Systems & Control Letters 16 (1991) 309–317.

[6] J.C. Willems, Dissipative dynamical systems. Part I: general theory, Archive for Rational Mechanics and Analysis 45 (1992) 321–351.

[7] D.J. Hill, P.J. Moylan, Dissipative dynamical systems: basic input–output and state properties, Journal of the Franklin Institute 309 (1980) 327–357.

[8] F. Rasool, D. Huang, S.K. Nguang, Robust H_∞ output feedback control of networked control systems with multiple quantizers, Journal of the Franklin Institute 349 (3) (2012) 1153–1173.

[9] S.K. Nguang, W. Assawinchaichote, P. Shi, Y. Shi, Robust H_∞ control design for uncertain fuzzy systems with Markovian jumps: an LMI approach, in: American Control Conference, 2005, pp. 1805–1810.

[10] S.K. Nguang, P. Zhang, S.X. Ding, Parity relation based fault estimation for nonlinear systems: an LMI approach, International Journal of Automation and Computing 4 (2) (2007) 164–168.
[11] S.K. Nguang, Comments on "Robust stabilization of uncertain input-delay systems by sliding mode control with delay compensation", Automatica 37 (10) (2001) 1677.
[12] S.K. Nguang, P. Shi, H_∞ output feedback control of fuzzy system models under sampled measurements, Computers and Mathematics With Applications 46 (5) (2003) 705–717.
[13] W. Assawinchaichote, S.K. Nguang, P. Shi, Fuzzy Control and Filter Design for Uncertain Fuzzy Systems, Springer, 2006.
[14] S.K. Nguang, P. Shi, On designing filters for uncertain sampled-data nonlinear systems, Systems & Control Letters 41 (5) (2000) 305–316.
[15] D. Huang, S.K. Nguang, Robust Control for Uncertain Networked Control Systems With Random Delays, Springer Science & Business Media, 2009.
[16] J. Zhang, A.K. Swain, S.K. Nguang, Robust sensor fault estimation scheme for satellite attitude control systems, Journal of the Franklin Institute 350 (9) (2013) 2581–2604.
[17] J. Zhang, P. Shi, J. Qiu, S.K. Nguang, A novel observer-based output feedback controller design for discrete-time fuzzy systems, IEEE Transactions on Fuzzy Systems 23 (1) (2015) 223–229.
[18] S.K. Nguang, P. Shi, Delay-dependent H_∞ filtering for uncertain time delay nonlinear systems: an LMI approach, IET Control Theory & Applications 1 (1) (2007) 133–140.
[19] Z. Hou, J. Luo, P. Shi, S.K. Nguang, Stochastic stability of Ito differential equations with semi-Markovian jump parameters, IEEE Transactions on Automatic Control 51 (8) (2006) 1383–1387.
[20] W. Assawinchaichote, S.K. Nguang, P. Shi, E.K. Boukas, H_∞ fuzzy state-feedback control design for nonlinear systems with stability constraints: an LMI approach, Mathematics and Computers in Simulation 78 (4) (2008) 514–531.
[21] S. Chae, S.K. Nguang, SOS based robust H_∞ fuzzy dynamic output feedback control of nonlinear networked control systems, IEEE Transactions on Cybernetics 44 (7) (2014) 1204–1213.
[22] F. Rasool, D. Huang, S.K. Nguang, Robust H_∞ output feedback control of discrete-time networked systems with limited information, Systems & Control Letters 60 (10) (2011) 845–853.
[23] S.K. Nguang, P. Shi, Stabilisation of a class of nonlinear time-delay systems using fuzzy models, in: Proceedings of the 39th IEEE Conference on Decision and Control, 2000, pp. 5–11.
[24] S. Saat, S.K. Nguang, Nonlinear H_∞ output feedback control with integrator for polynomial discrete-time systems, International Journal of Robust and Nonlinear Control 25 (2015) 1051–1065.
[25] Y. Zhang, P. Shi, S.K. Nguang, H.R. Karimi, Observer-based finite-time fuzzy H_∞ control for discrete-time systems with stochastic jumps and time-delays, Signal Processing 97 (2014) 252–261.
[26] S. Chae, F. Rasool, S.K. Nguang, A. Swain, Robust mode delay-dependent H_∞ control of discrete-time systems with random communication delays, IET Control Theory & Applications 4 (6) (2010) 936–944.

[27] Y. Zhang, P. Shi, S.K. Nguang, Y. Song, Robust finite-time H_∞ control for uncertain discrete-time singular systems with Markovian jumps, IET Control Theory & Applications 8 (12) (2014) 1105–1111.

[28] P.A. Parrilo, Structured Semidefinite Programs and Semialgebraic Geometry Methods in Robustness and Optimization, PhD dissertation, California Inst. Technol., Pasadena, 2000.

[29] S. Prajna, A. Papachristodoulou, P. Seiler, SOSTOOLS: Sum of Squares Optimization Toolbox for MATLAB, User's Guide, 2004.

[30] A. Papachristodoulou, S. Prajna, On the construction of Lyapunov functions using the sum of squares decomposition, in: Proceedings of the 41st IEEE Conference on Decision and Control (CDC), Las Vegas, 2002.

[31] S. Prajna, A. Papachristodoulou, F. Wu, Nonlinear control synthesis by sum of squares optimization: a Lyapunov-based approach, in: Proceedings of the 5th Asian Control Conference, 2004, pp. 157–165.

[32] H.J. Ma, G.H. Yang, Fault-tolerant control synthesis for a class of nonlinear systems: sum of squares optimization approach, International Journal of Robust and Nonlinear Control 19 (5) (2009) 591–610.

[33] D. Zhao, J. Wang, An improved H_∞ synthesis for parameter-dependent polynomial nonlinear systems using sos programming, in: American Control Conference, 2009, pp. 796–801.

[34] H.J. Ma, G.H. Yang, Fault tolerant H_∞ control for a class of nonlinear discrete-time systems: using sum of squares optimization, in: Proceeding of American Control Conference, 2008, pp. 1588–1593.

[35] H.K. Khalil, Nonlinear Systems, Prentice-Hall, 1996.

[36] S.K. Nguang, Robust nonlinear H_∞ output feedback control, IEEE Transactions on Automatic Control 4 (7) (July 1996) 1003–1007.

[37] S.K. Nguang, W. Assawinchaichote, H_∞ filtering for fuzzy dynamical systems with D-stability constraints, IEEE Transactions on Circuits and Systems. I, Regular Papers 50 (11) (Nov. 2003) 1503–1508.

Chapter 4

Robust nonlinear filtering for polynomial discrete-time systems

CHAPTER OUTLINE

4.1 INTRODUCTION

We consider the availability of all the states for direct measurement as a rare occasion in practical feedback control systems. Besides, in most cases, there is a need for a reliable estimation of unmeasurable state variables [1–25]. The reliable estimation is needed especially for the synthesis of model-based controllers or for process monitoring purposes. For this purpose, a filter is usually employed to accurately estimate the state. The results of filter designs for nonlinear systems can be found in [26–28]. In particular, in [26], a solution in terms of the Hamilton–Jacobi inequalities (HJIs) is proposed. It is well known that solving the HJIs is hard because there are no computational tools available for solving them. Meanwhile, a convex solution to the filter problem has been given in [27] through the S-procedure method. In this paper, the problem is formulated in LMI forms and solved using LMI toolbox. However, to render a convex solution, some assumptions about nonlinear terms of the error dynamics have to be made. The assumptions may cause the results to be conservative. On the other hand, the filter design for polynomial systems have been considered in [28]. The author has shown that a convex solution can be rendered without requiring any

Analysis and Synthesis of Polynomial Discrete-Time Systems. DOI: 10.1016/B978-0-08-101901-6.00004-9

assumptions about nonlinear terms of the error dynamics. By utilizing the SOS programming approach the convex problem can be solved efficiently. Unfortunately, the result proposed in this paper is only held locally.

In this chapter, we attempt to design a nonlinear filter to estimate the state of polynomial discrete-time systems with and without uncertainties. A nonlinear filter design method for polynomial discrete-time systems with and without uncertainties by using SOS-SDP is established without any assumptions on nonlinear terms of the error dynamics. To ensure that a convex solution to the filter design problem can be obtained, an integrator is incorporated into the filter structure. To compute the filter gains, SOS techniques have been used to reduce the problems to SDP. The effectiveness of the proposed method is confirmed through simulation examples.

4.2 SYSTEM DESCRIPTION AND DEFINITION

Consider the following polynomial discrete-time system:

$$\begin{aligned} x(k+1) &= A(x(k))x(k), \\ y &= C(x(k))x(k), \end{aligned} \tag{4.1}$$

where $x(k) \in \Re^n$ is the state, y is the measurement, and $A(x(k))$ and $C(x(k))$ are polynomial matrices of appropriate dimensions.

A filter to estimate the state $x(k)$ from y is selected to be of the following form:

$$\begin{aligned} \hat{x}(k+1) =& A(\hat{x}(k))\hat{x} + L(\hat{x}(k))\big(y - \hat{y}\big), \\ \hat{y} =& C(\hat{x}(k))\hat{x}(k), \end{aligned} \tag{4.2}$$

where $\hat{x}$ is a filter state, $\hat{y}$ is a filter measurement, and $L(\hat{x}(k))$ is a designed polynomial matrix with appropriate dimensions.

To study the convergence and performance of the filter (4.2), we will look at the dynamics of the estimation error defined by $e = \hat{x}(k) - x(k)$. The error dynamics is given as follows:

$$\begin{aligned} e(k+1) &= \hat{x}(k+1) - x(k+1) \\ &= A(\hat{x}) + L(\hat{x})\big(C(\hat{x})\hat{x} - C(x)x\big) - A(x)x \\ &= \big[A(\hat{x}) + L(\hat{x})C(\hat{x})\big]e \\ &\quad + \big[A(\hat{x}) - A(x) + L(\hat{x})C(\hat{x}) - L(\hat{x})C(x)\big]x. \end{aligned} \tag{4.3}$$

Now, letting $\tilde{e} = [e^T, x^T]^T$, system (4.3) can be rewritten as

$$\tilde{e}(k+1) = \phi(x, \hat{x})\tilde{e}, \tag{4.4}$$

where

$$\phi(x, \hat{x}) = \begin{bmatrix} A(\hat{x}) + L(\hat{x})C(\hat{x}) & A(\hat{x}) - A(x) + L(\hat{x})C(\hat{x}) - L(\hat{x})C(x) \\ 0 & A(x) \end{bmatrix}. \tag{4.5}$$

Theorem 9. *Consider system* (4.1)*. The error dynamics shown in* (4.4) *is asymptotically stable if there exist polynomial matrices $L(\hat{x})$ and $P(\tilde{e})$ such that the following conditions are satisfied:*

$$P(\tilde{e}) > 0, \tag{4.6}$$

$$\begin{bmatrix} P(\tilde{e}) & \phi^T(x, \hat{x})P^T(\tilde{e}(k+1)) \\ P(\tilde{e}(k+1))\phi(x, \hat{x}) & P(\tilde{e}(k+1)) \end{bmatrix} > 0. \tag{4.7}$$

Proof. The parameter-dependent Lyapunov function is selected as follows:

$$V(\tilde{e}) = \tilde{e}^T P(\tilde{e})\tilde{e}. \tag{4.8}$$

The difference between $V(\tilde{e}(k+1))$ and $V(\tilde{e}(k))$ along (4.4) with (4.2) is given by

$$\begin{aligned} \Delta(V(\tilde{e})) &= V(\tilde{e}(k+1)) - V(\tilde{e}) \\ &= \tilde{e}^T(k+1)P(\tilde{e}(k+1))\tilde{e}(k+1) - \tilde{e}^T P(\tilde{e})\tilde{e} \\ &= \tilde{e}^T\big[\phi^T(x, \hat{x})P(\tilde{e}(k+1))\phi(x, \hat{x}) - P(\tilde{e})\big]\tilde{e}. \end{aligned} \tag{4.9}$$

Suppose (4.7) is feasible. Multiplying it to the left by $diag[I, P(\tilde{e}(k+1))]$ and to the right by $diag[I, P^T(\tilde{e}(k+1))]$ and then applying the Schur complement, we have

$$\phi^T(x, \hat{x})P(\tilde{e}(k+1))\phi(x, \hat{x}) - P(\tilde{e}) < 0. \tag{4.10}$$

Knowing that (4.10) holds, we have $\Delta V(\tilde{e}) < 0$, which implies that the error dynamics (4.4) with the filter (4.2) is globally asymptotically stable. □

Remark 21. Theorem 9 provides a sufficient condition for the existence of filter gains and is given in terms of solutions to a set of parameterized PMIs. However, notice that the $P(\tilde{e}(k+1))$ appears in the PMIs, and therefore the inequalities are not jointly convex. We might choose to select the

Lyapunov matrix to be of $P(e)$ instead of $P(\tilde{e})$. However, such a selection does not help the solution to be convex because the problem remains persistent. Hence, to directly solve Theorem 9 is hard because the PMIs need to be checked for all combinations of $P(\tilde{e})$ and $L(\hat{x})$, which results in solving an infinite number of PMIs. In light of the aforementioned problem, in our work, we propose to incorporate an integrator into the filter dynamics. In doing so, a convex solution to the filter design problem for polynomial discrete-time systems can be rendered efficiently. The details of this integrator method are illustrated in the following section.

4.3 MAIN RESULTS

In this section, the significance of incorporating an integrator into the filter dynamics is illustrated for polynomial discrete-time systems with or without uncertainties.

4.3.1 Nonlinear filtering for polynomial discrete-time systems

The nonlinear filter with an integrator is given as

$$\begin{aligned} \hat{x}(k+1) &= A(\hat{x})\hat{x} + x_f, \\ x_f(k+1) &= x_f + L(\hat{x})\big(C(\hat{x})\hat{x} - C(x)x\big), \end{aligned} \tag{4.11}$$

where $\hat{x}$ is a filter state, x_f is an augmented filter state, $L(\hat{x})$ is a designed polynomial matrix, and $A(\hat{x})$, $C(\hat{x})$, and $C(x)$ are all polynomial matrices of appropriate dimensions.

Now, the estimation error is defined as

$$\bar{e} = \begin{bmatrix} e_1 \\ e_2 \end{bmatrix} = \begin{bmatrix} \hat{x} - x \\ x_f \end{bmatrix}. \tag{4.12}$$

Then estimation error dynamics is given by

$$\begin{aligned} \bar{x}(k+1) &= \begin{bmatrix} e_1(k+1) \\ e_2(k+1) \end{bmatrix} = \begin{bmatrix} \hat{x}(k+1) - x(k+1) \\ x_f(k+1) \end{bmatrix} \\ &= \begin{bmatrix} A(\hat{x})\hat{x} + x_f - A(x)x \\ x_f + L(\hat{x})\big(C(\hat{x})\hat{x} - C(x)x\big) \end{bmatrix} \\ &= \begin{bmatrix} A(\hat{x})e_1 + e_2 + \big(A(\hat{x}) - A(x)\big)x \\ e_2 + L(\hat{x})C(\hat{x})e_1 + \big(L(\hat{x})C(\hat{x}) - L(\hat{x})C(x)\big)x \end{bmatrix}. \end{aligned} \tag{4.13}$$

Next, defining $\check{e} = [e_1^T, x^T, e_2^T]^T$, the estimation error dynamics described in (4.13) can be rewritten as

$$\check{e}(k+1) = \phi_2(x, \hat{x})\check{e}, \tag{4.14}$$

where

$$\phi_2(x, \hat{x}) = \begin{bmatrix} A(\hat{x}) & A(\hat{x}) - A(x) & 1 \\ 0 & A(x) & 0 \\ L(\hat{x})C(\hat{x}) & L(\hat{x})C(\hat{x}) - L(\hat{x})C(x) & 1 \end{bmatrix}. \tag{4.15}$$

Theorem 10. *Consider system* (4.1)*. The error dynamics shown in* (4.14) *is asymptotically stable if there exist a symmetric polynomial matrix* $P(e_1)$ *and polynomial matrices* $K(\hat{x})$ *and* $G(\hat{x})$ *such that the following conditions are satisfied:*

$$P(e_1) > 0, \tag{4.16}$$

$$\begin{bmatrix} P(e_1) & \phi_2^T(x, \hat{x})G^T(\hat{x}) \\ G(\hat{x})\phi_2(x, \hat{x}) & G^T(\hat{x}) + G(\hat{x}) - P(e_1(k+1)) \end{bmatrix} > 0, \tag{4.17}$$

where

$$G(\hat{x}) = \begin{bmatrix} G_{11}(\hat{x}) & G_{12}(\hat{x}) & G_{13}(\hat{x}) \\ G_{21}(\hat{x}) & G_{22}(\hat{x}) & G_{13}(\hat{x}) \\ G_{31}(\hat{x}) & G_{32}(\hat{x}) & G_{13}(\hat{x}) \end{bmatrix}. \tag{4.18}$$

Moreover, the nonlinear filter is

$$\begin{aligned} \hat{x}(k+1) &= A(\hat{x})\hat{x} + x_f, \\ x_f(k+1) &= x_f + L(\hat{x})\big(C(\hat{x})\hat{x} - C(x)x\big), \end{aligned} \tag{4.19}$$

where

$$L(\hat{x}) = K(\hat{x})G_{12}^{-1}(\hat{x}). \tag{4.20}$$

Proof. We select the following Lyapunov:

$$V(\check{e}) = \check{e}^T P(e_1)\check{e}. \tag{4.21}$$

Then, the difference between $V(\check{e}(k+1))$ and $V(\check{e}(k))$ along (4.14) with (4.11) is given by

$$\begin{aligned} \Delta(V(\check{e})) &= V(\check{e}(k+1)) - V(\check{e}) \\ &= \check{e}^T(k+1)P(e_1(k+1))\check{e}(k+1) - \check{e}^T P(e_1)\check{e} \\ &= \check{e}^T\big[\phi_2^T(x, \hat{x})P(e_1(k+1))\phi_2(x, \hat{x}) - P(\check{e})\big]\check{e}. \end{aligned} \tag{4.22}$$

Suppose (4.17) is feasible. Then $G^T(\hat{x}) + G(\hat{x}) > P(e_1(k+1)) > 0$. This implies that $G(\hat{x})$ is nonsingular. Since $P(e_1(k+1))$ is positive definite, we get the inequality

$$\big(P(e_1(k+1)) - G(\hat{x})\big)P^{-1}(e_1(k+1))\big(P(e_1(k+1)) - G(\hat{x})\big)^T > 0 \tag{4.23}$$

or

$$G(\hat{x}(k))P^{-1}(e_1(k+1))G^T(\hat{x}) \geq G(\hat{x}) + G^T(\hat{x}) - P(e_1(k+1)). \tag{4.24}$$

This immediately gives

$$\begin{bmatrix} P(e_1) & \phi_2^T(x,\hat{x})G^T(\hat{x}) \\ G(\hat{x})\phi_2(x,\hat{x}) & G^T(\hat{x})P^{-1}(e_1(k+1))G(\hat{x}) \end{bmatrix} > 0. \tag{4.25}$$

Next, multiplying (4.25) on the right by $diag[I, G^{-1}(\hat{x}(k))]^T$ and on the left by $diag[I, G^{-1}(\hat{x}(k))]$, we get

$$\begin{bmatrix} P(e_1) & \phi_2^T(x,\hat{x}) \\ \phi_2(x,\hat{x}) & P^{-1}(e_1(k+1)) \end{bmatrix} > 0. \tag{4.26}$$

Then, applying the Schur complement into (4.26), we have

$$\phi_2^T(x,\hat{x})P(e_1(k+1))\phi_2(x,\hat{x}) - P(e_1) < 0. \tag{4.27}$$

Knowing that (4.27) holds, we have $\Delta V(\check{e}) < 0$, which implies that the error dynamics (4.14) with the filter (4.11) is globally asymptotically stable. The proof ends. □

Remark 22. We might wonder how the term $L(\hat{x}) = K(\hat{x})G_2^{-1}(\hat{x})$ can suddenly appear in Theorem 10. The fact is that a change-of-variable technique has been applied in the above proof, where $K(\hat{x}) = L(\hat{x})G_{12}(\hat{x})$. This is explicitly applied in Theorem 10. It is also important to note that to allow the same value of $L(\hat{x})$ to be obtained, the polynomial matrix $G(\hat{x})$ must be enforced to be of a certain structure; see Eq. (4.18). Although $G(\hat{x})$ must be of a certain form, the results are still not too conservative because it is independent of the Lyapunov matrix.

Remark 23. Inequalities (4.17) of Theorem 10 are convex. This is true because the terms in $P(e_1(k+1))$ are jointly convex. For clarity, refer to the following expansion version of $P(e_1(k+1))$:

$$P(e_1(k+1)) = P\big[A(\hat{x})\hat{x} + x_f - A(x)x\big]. \tag{4.28}$$

From (4.28), the x_f is an augmented state, and hence $P(e_1(k+1))$ provides a convex solution. Therefore, Theorem 10 can possibly be solved via SDP.

Unfortunately, solving Theorem 10 is hard because we need to solve an infinite set of state-dependent PMIs. To relax these conditions, we utilize an SOS decomposition approach [29], and therefore the conditions given in Theorem 10 can be converted into SOS conditions, which are given by the following corollary.

Corollary 4. *Consider system* (4.1)*. The error dynamics shown in* (4.14) *is asymptotically stable if there exist a symmetric polynomial matrix* $P(e_1)$*, polynomial matrices* $K(\hat{x})$ *and* $G(\hat{x})$*, and positive constants* ϵ_1 *and* ϵ_2 *such that the following conditions are satisfied:*

$$v_1^T[P(e_1)-\epsilon_1 I]v_1 \quad \text{is an SOS}, \tag{4.29}$$

$$v_2^T\begin{bmatrix} P(e_1)-\epsilon_2 I & \phi_2^T(x,\hat{x})G^T(\hat{x}) \\ G(\hat{x})\phi_2(x,\hat{x}) & G^T(\hat{x})+G(\hat{x})-P(e_1(k+1))-\epsilon_2 I \end{bmatrix} v_2 \quad \text{is an SOS}, \tag{4.30}$$

where, v_1 *and* v_2 *are free vectors of appropriate dimensions, and*

$$G(\hat{x}) = \begin{bmatrix} G_{11}(\hat{x}) & G_{12}(\hat{x}) & G_{13}(\hat{x}) \\ G_{21}(\hat{x}) & G_{22}(\hat{x}) & G_{13}(\hat{x}) \\ G_{31}(\hat{x}) & G_{32}(\hat{x}) & G_{13}(\hat{x}) \end{bmatrix}. \tag{4.31}$$

Furthermore, the nonlinear filter is given by

$$\begin{aligned} \hat{x}(k+1) &= A(\hat{x})\hat{x} + x_f, \\ x_f(k+1) &= x_f + L(\hat{x})\big(C(\hat{x})\hat{x} - C(x)x\big), \end{aligned} \tag{4.32}$$

where

$$L(\hat{x}) = K(\hat{x})G_{12}^{-1}(\hat{x}). \tag{4.33}$$

Remark 24. It should be mentioned that the above design procedures are dedicated to solve the full-order filter design problems. We only present the most fundamental filter design procedure without inclusion of any performance objectives or uncertainties. The above idea only provides a possible solution to the global filter design problem of polynomial discrete-time systems. Some conservatism of the proposed method is given as follows:

- The slack polynomial matrix $G(\hat{x})$ must be of a certain structure to achieve a feasible solution of the problem. Although this conservativeness is not very severe because the slack polynomial matrix is independent of the Lyapunov matrix, it is still conservative in general.
- The computational complexity must be one of the important issues using this method. This is because the filter state proposed in this method

has double of the order of an original state. Therefore, if we are dealing with the second-order system, then the filter must be of order four. Consequently, it will give burden to the computational aspects. The computational complexity becomes even worse because all matrices are in polynomial forms. Therefore, this method can only be applied to low-order academic examples.

4.3.2 Robust nonlinear filtering for polynomial discrete-time systems

Consider the following polynomial discrete-time system with parametric uncertainties:

$$\begin{aligned} x(k+1) &= A_o(x(k))x(k) + A(x(k), \theta)x(k), \\ y &= C_o(x(k))x(k) + C(x(k), \theta)x(k), \end{aligned} \tag{4.34}$$

where $A_o(x(k))$ and $C_o(x(k))$ are matrix functions, and the matrix functions $\cdot(x(k), \theta)$ are defined as

$$A(x(k), \theta) = \sum_{i=1}^{q} A_i(x(k))\theta_i. \tag{4.35}$$

$\theta = \left[\theta_1, \dots, \theta_q\right]^T \in \mathbb{R}^q$ is the vector of constant uncertainty and satisfies

$$\theta \in \Theta \triangleq \left\{ \theta \in \mathbb{R}^q : \theta_i \geq 0, i = 1, \dots, q, \sum_{i=1}^{q} \theta_i = 1 \right\}. \tag{4.36}$$

The robust filter is given by

$$\begin{aligned} \hat{x}(k+1) &= A_o(\hat{x})\hat{x} + x_f, \\ x_f(k+1) &= x_f + L(\hat{x})\big(C_o(\hat{x})\hat{x} - C(x)x\big). \end{aligned} \tag{4.37}$$

Define the error as follows:

$$\bar{e} = \begin{bmatrix} e_1 \\ e_2 \end{bmatrix} = \begin{bmatrix} \hat{x} - x \\ x_f \end{bmatrix}. \tag{4.38}$$

The error dynamics is then given by

$$\bar{x}(k+1) = \begin{bmatrix} e_1(k+1) \\ e_2(k+1) \end{bmatrix} = \begin{bmatrix} \hat{x}(k+1) - x(k+1) \\ x_f(k+1) \end{bmatrix}$$

$$
\begin{aligned}
&= \begin{bmatrix} A_o(\hat{x})\hat{x} + x_f - A_o(x)x - A(x,\theta)x \\ x_f + L(\hat{x})\big(C_o(\hat{x})\hat{x} - C_o(x) - C(x,\theta)x\big) \end{bmatrix} \\
&= \begin{bmatrix} A_o(\hat{x})e_1 + e_2 + \big(A_o(\hat{x}) - A_o(x) - A(x,\theta)\big)x \\ e_2 + L(\hat{x})C_o(\hat{x})e_1 + \big(L(\hat{x})C(\hat{x}) - L(\hat{x})[C_o(x) + C(x,\theta)]\big)x \end{bmatrix}.
\end{aligned}
\tag{4.39}
$$

Next, defining $\check{e} = [e_1, x, e_2]^T$, the error dynamics described in (4.39) can be rewritten as

$$
\check{e}(k+1) = \sum_{i=1}^{q} \theta_i \Phi_i(x,\hat{x})\check{e}, \tag{4.40}
$$

where

$$
\Phi_i(x,\hat{x}) = \begin{bmatrix} A_o(\hat{x}) & A_o(\hat{x}) - A_o(x) - A_i(x) & 1 \\ 0 & A_o(x) + A_i(x) & 0 \\ L(\hat{x})C_o(\hat{x}) & L(\hat{x})C_o(\hat{x}) - L(\hat{x})[C_o(x) + C_i(x)]) & 1 \end{bmatrix}. \tag{4.41}
$$

Theorem 11. *Consider system* (4.34). *The error dynamics shown in* (4.40) *is asymptotically stable if there exist symmetric polynomial matrices* $P_i(e_1)$ *and polynomial matrices* $K(\hat{x})$ *and* $G(\hat{x})$ *such that the following conditions are satisfied for* $i = 1, \ldots, q$:

$$
P_i(e_1) > 0, \tag{4.42}
$$

$$
\begin{bmatrix} P_i(e_1) & \Phi_i^T(x,\hat{x})G^T(\hat{x}) \\ G(\hat{x})\Phi_i(x,\hat{x}) & G^T(\hat{x}) + G(\hat{x}) - P_i(e_1(k+1)) \end{bmatrix} > 0, \tag{4.43}
$$

where

$$
G(\hat{x}) = \begin{bmatrix} G_{11}(\hat{x}) & G_{12}(\hat{x}) & G_{13}(\hat{x}) \\ G_{21}(\hat{x}) & G_{22}(\hat{x}) & G_{13}(\hat{x}) \\ G_{31}(\hat{x}) & G_{32}(\hat{x}) & G_{13}(\hat{x}) \end{bmatrix}. \tag{4.44}
$$

Moreover, the nonlinear filter is given by

$$
\begin{aligned}
\hat{x}(k+1) &= A_o(\hat{x})\hat{x} + x_f, \\
x_f(k+1) &= x_f + L(\hat{x})\big(C_o(\hat{x})\hat{x} - y\big),
\end{aligned}
\tag{4.45}
$$

where

$$
L(\hat{x}) = K(\hat{x})G_{12}^{-1}(\hat{x}). \tag{4.46}
$$

Proof. Choosing the Lyapunov function

$$V(\check{e}) = \check{e}^T \sum_{i=1}^{q} \theta_i P_i(e_1)\check{e}, \tag{4.47}$$

the theorem can be proven in the same way as Theorem 10. □

Similarly, Theorem 11 can be converted into SOS conditions given by the following corollary.

Corollary 5. *Consider system* (4.34). *The error dynamics shown in* (4.40) *is asymptotically stable if there exist a symmetric polynomial matrix* $P_i(e_1)$, *polynomial matrices* $K(\hat{x})$ *and* $G(\hat{x})$, *and positive constants* ϵ_1 *and* ϵ_2 *such that the following conditions are satisfied for* $i = 1, \ldots, q$*:*

$$v_1^T[P_i(e_1) - \epsilon_1 I]v_1 \quad \text{is an SOS,} \tag{4.48}$$

$$v_2^T \begin{bmatrix} P_i(e_1) - \epsilon_2 I & \Phi_i^T(x,\hat{x})G^T(\hat{x}) \\ G(\hat{x})\Phi_i(x,\hat{x}) & G^T(\hat{x}) + G(\hat{x}) - P_i(e_1(k+1)) - \epsilon_2 I \end{bmatrix} v_2 \quad \text{is an SOS,} \tag{4.49}$$

where v_1 *and* v_2 *are free vectors of appropriate dimensions, and*

$$G(\hat{x}) = \begin{bmatrix} G_{11}(\hat{x}) & G_{12}(\hat{x}) & G_{13}(\hat{x}) \\ G_{21}(\hat{x}) & G_{22}(\hat{x}) & G_{13}(\hat{x}) \\ G_{31}(\hat{x}) & G_{32}(\hat{x}) & G_{13}(\hat{x}) \end{bmatrix}. \tag{4.50}$$

Furthermore, the nonlinear filter is given by

$$\begin{aligned} \hat{x}(k+1) &= A_o(\hat{x})\hat{x} + x_f, \\ x_f(k+1) &= x_f + L(\hat{x})\big(C_o(\hat{x})\hat{x} - y\big), \end{aligned} \tag{4.51}$$

where

$$L(\hat{x}) = K(\hat{x})G_{12}^{-1}(\hat{x}). \tag{4.52}$$

4.4 NUMERICAL EXAMPLES

Example 1. Consider the polynomial discrete system

$$\begin{aligned} x(k+1) &= \begin{bmatrix} 1 & -0.01 \\ 0.01 + 0.01x_1x_2 & 1 - 0.01x_2 \end{bmatrix} \begin{bmatrix} x_1 \\ x_2 \end{bmatrix}, \\ y &= x_1. \end{aligned} \tag{4.53}$$

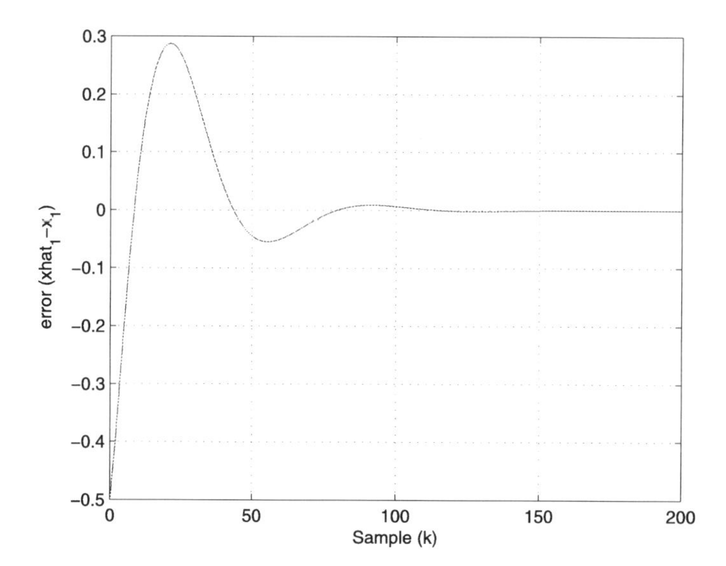

■ FIGURE 4.1 Trajectory of the $\hat{x}_1 - x_1$.

Then, by applying Corollary 10 with $P(e_1)$ of degree 2, the polynomial matrix $G(\hat{x})$ of degree 4, and polynomial matrix $K(\hat{x})$ of degree of 6 we obtain a feasible solution. The results of the error between the estimation state and the actual state can be seen in Figs. 4.1 and 4.2. The initial condition for the actual state is $x(0) = [1 \quad 1]$, and for the filter state, it is $\hat{x}(0) = [0.5 \quad 0.5]$.

Remark 25. The values of the polynomial matrices $P(e_1)$, $G(\hat{x})$, and $K(\hat{x})$ are omitted here due to their large sizes.

Example 2. Consider the polynomial discrete system with parametric uncertainties of the form (4.34) with

$$A_o(x) = \begin{bmatrix} 1 + 0.1x_1 & -0.1 \\ 0.1 + 0.1x_1x_2 & 1 - 0.1x_2 \end{bmatrix},$$

$$A_1(x) = \begin{bmatrix} 0 & -0.01x_1 \\ 0.01x_1 & -0.01x_2 \end{bmatrix}, \quad A_2(x) = \begin{bmatrix} 1 & 0 \\ 0.01x_1x_2 & 0 \end{bmatrix},$$

and

$$C_o(x) = C_1(x) = C_2(0) = [1; \ 0].$$

Applying Corollary 5 with $P(e_1)$ of degree 5, the polynomial matrix $G(\hat{x})$ of degree 6, and the polynomial matrix $K(\hat{x})$ of degree 8, we obtain a feasible solution. The estimation errors are depicted in Figs. 4.3 and 4.4. The

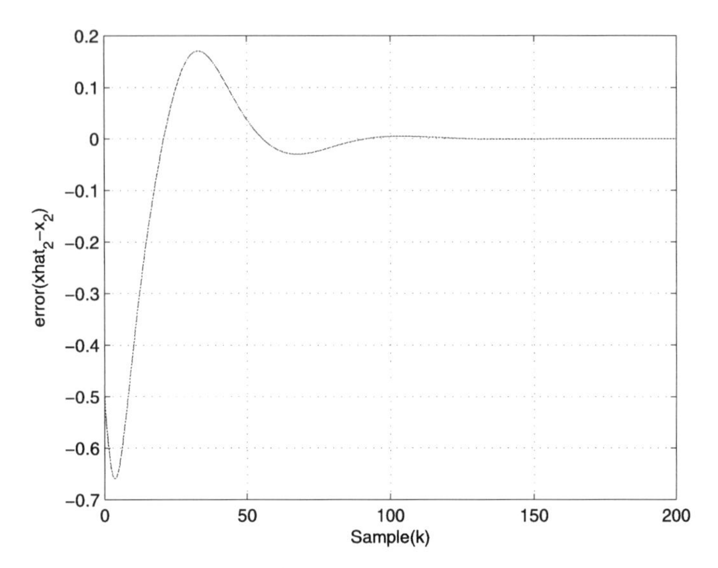

■ **FIGURE 4.2** Trajectory of the $\hat{x}_2 - x_2$.

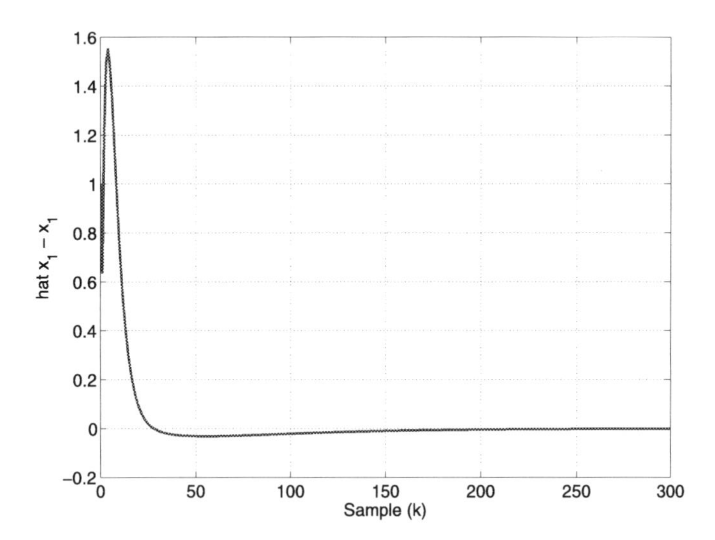

■ **FIGURE 4.3** Trajectory of the $\hat{x}_1 - x_1$.

initial condition for the actual state is $x(0) = [1 \quad 1]$, and for the filter state, it is $\hat{x}(0) = [0 \quad 0]$. We can see that the nonlinear filter is able to estimate the plant states.

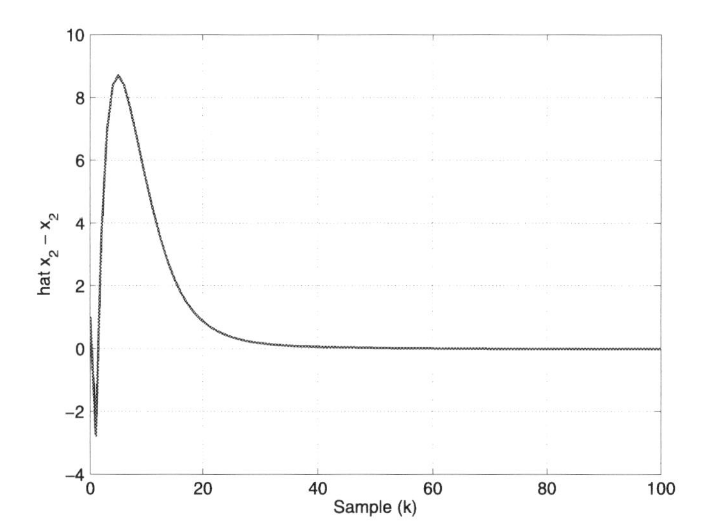

■ **FIGURE 4.4** Trajectory of the $\hat{x}_2 - x_2$.

4.5 CONCLUSION

The robust nonlinear filter design problem for polynomial discrete-time systems with and without uncertainties has been examined. It has been shown that a convex solution to the problem can be obtained efficiently by incorporating an integrator into the filter structure. Sufficient conditions for the existence of a robust nonlinear filter are given in terms of the solvability of the PMIs, which have been formulated as SOS constraints and can be solved by any SOS solver.

REFERENCES

[1] J.L. Anderson, An ensemble adjustment Kalman filter for data assimilation, Monthly Weather Review 129 (2001) 2884–2903.

[2] R.E. Kalman, A new approach to linear filtering and prediction problems, Journal of Basic Engineering 82 (1960) 35–45.

[3] M. Fisher, M. Leutbecher, G.A. Kelly, On the equivalence between Kalman smoothing and weak-constraint four-dimensional variational data assimilation, Quarterly Journal of the Royal Meteorological Society 131 (2005) 3235–3246.

[4] A.H. Jazwinski, Stochastic Processes and Filtering Theory, Academic Press, 1998.

[5] F. Rasool, D. Huang, S.K. Nguang, Robust H_∞ output feedback control of networked control systems with multiple quantizers, Journal of the Franklin Institute 349 (3) (2012) 1153–1173.

[6] S.K. Nguang, W. Assawinchaichote, P. Shi, Y. Shi, Robust H_∞ control design for uncertain fuzzy systems with Markovian jumps: an LMI approach, in: American Control Conference, 2005, pp. 1805–1810.

[7] S.K. Nguang, P. Zhang, S.X. Ding, Parity relation based fault estimation for nonlinear systems: an LMI approach, International Journal of Automation and Computing 4 (2) (2007) 164–168.
[8] S.K. Nguang, Comments on "Robust stabilization of uncertain input-delay systems by sliding mode control with delay compensation", Automatica 37 (10) (2001) 1677.
[9] S.K. Nguang, P. Shi, H_∞ output feedback control of fuzzy system models under sampled measurements, Computers and Mathematics With Applications 46 (5) (2003) 705–717.
[10] W. Assawinchaichote, S.K. Nguang, P. Shi, Fuzzy Control and Filter Design for Uncertain Fuzzy Systems, Springer, 2006.
[11] S.K. Nguang, P. Shi, On designing filters for uncertain sampled-data nonlinear systems, Systems & Control Letters 41 (5) (2000) 305–316.
[12] D. Huang, S.K. Nguang, Robust Control for Uncertain Networked Control Systems With Random Delays, Springer Science & Business Media, 2009.
[13] J. Zhang, A.K. Swain, S.K. Nguang, Robust sensor fault estimation scheme for satellite attitude control systems, Journal of the Franklin Institute 350 (9) (2013) 2581–2604.
[14] J. Zhang, P. Shi, J. Qiu, S.K. Nguang, A novel observer-based output feedback controller design for discrete-time fuzzy systems, IEEE Transactions on Fuzzy Systems 23 (1) (2015) 223–229.
[15] S.K. Nguang, P. Shi, Delay-dependent H_∞ filtering for uncertain time delay nonlinear systems: an LMI approach, IET Control Theory & Applications 1 (1) (2007) 133–140.
[16] Z. Hou, J. Luo, P. Shi, S.K. Nguang, Stochastic stability of Ito differential equations with semi-Markovian jump parameters, IEEE Transactions on Automatic Control 51 (8) (2006) 1383–1387.
[17] W. Assawinchaichote, S.K. Nguang, P. Shi, E.K. Boukas, H_∞ fuzzy state-feedback control design for nonlinear systems with-stability constraints: an LMI approach, Mathematics and Computers in Simulation 78 (4) (2008) 514–531.
[18] S. Chae, S.K. Nguang, SOS based robust H_∞ fuzzy dynamic output feedback control of nonlinear networked control systems, IEEE Transactions on Cybernetics 44 (7) (2014) 1204–1213.
[19] F. Rasool, D. Huang, S.K. Nguang, Robust H_∞ output feedback control of discrete-time networked systems with limited information, Systems & Control Letters 60 (10) (2011) 845–853.
[20] S.K. Nguang, P. Shi, Stabilisation of a class of nonlinear time-delay systems using fuzzy models, in: Proceedings of the 39th IEEE Conference on Decision and Control, 2000, pp. 5–11.
[21] S. Saat, S.K. Nguang, Nonlinear H_∞ output feedback control with integrator for polynomial discrete-time systems, International Journal of Robust and Nonlinear Control 25 (2015) 1051–1065.
[22] Y. Zhang, P. Shi, S.K. Nguang, H.R. Karimi, Observer-based finite-time fuzzy H_∞ control for discrete-time systems with stochastic jumps and time-delays, Signal Processing 97 (2014) 252–261.
[23] S. Chae, F. Rasool, S.K. Nguang, A. Swain, Robust mode delay-dependent H_∞ control of discrete-time systems with random communication delays, IET Control Theory & Applications 4 (6) (2010) 936–944.

[24] Y. Zhang, P. Shi, S.K. Nguang, Y. Song, Robust finite-time H_∞ control for uncertain discrete-time singular systems with Markovian jumps, IET Control Theory & Applications 8 (12) (2014) 1105–1111.

[25] S.J. Julier, J.K. Uhlmann, New extension of the Kalman filter to nonlinear systems, in: I. Kader (Ed.), Signal Processing, Sensor Fusion, and Target Recognition VI, SPIE Proceedings, vol. 3068, International Society for Optical Engineering, 1997, pp. 182–193.

[26] S.K. Nguang, M. Fu, Robust nonlinear H_∞ filtering, Automatica 32 (8) (1996) 1195–1199.

[27] A. Howell, J.K. Hedrick, Nonlinear observer design via convex programming, in: Proceedings of American Control Conference, 2002, pp. 2088–2093.

[28] H. Ichihara, Observer design for polynomial systems using convex optimization, in: Proceedings of the 46th IEEE Conference on Decision and Control, 2007, pp. 5347–5352.

[29] P.A. Parrilo, Structured Semidefinite Programs and Semialgebraic Geometry Methods in Robustness and Optimization, PhD dissertation, California Inst. Technol., Pasadena, 2000.

Chapter 5

Robust nonlinear H_∞ filtering for polynomial discrete-time systems

CHAPTER OUTLINE

5.1 INTRODUCTION

H_∞ control and filtering for nonlinear systems has received considerable attention [1–31]. There are two commonly used approaches for providing solutions to nonlinear H_∞ problems. One is based on the dissipativity theory and theory of differential games (see [6] and [1]). The other is based on the nonlinear version of the classical bounded real lemma as developed by [7] and [8] (see, for example, [3,4], and [5]). Both approaches convert the problem of nonlinear H_∞ control and filtering to the solvability of the so-called Hamilton–Jacobi equation (HJE). A nice feature of these results is that they are parallel to the linear H_∞ results.

Further research along the line of the dissipativity theory and theory of differential games has been attempted, where results on disturbance attenuation for nonlinear systems via state feedback and/or output feedback have been provided. In [31] solutions to the nonlinear H_∞ filtering problem have been obtained, and the authors have established the equivalence between a robust nonlinear H_∞ filtering problem and nonlinear H_∞ filtering for a sys-

Analysis and Synthesis of Polynomial Discrete-Time Systems. DOI: 10.1016/B978-0-08-101901-6.00005-0

tem without uncertainty. This allows us to solve the robust nonlinear H_∞ filtering problem via existing nonlinear H_∞ filtering techniques [31].

Motivated by the fact that HJEs are very difficult to solve, in this chapter, we attempt to design an H_∞ filter for polynomial discrete-time systems with and without uncertainties. Based on the SOS approach, sufficient conditions for the existence of an H_∞ nonlinear filter for polynomial discrete-time systems are provided in terms of SOS constraints. Here, to ensure that a convex solution to the filter design problem can be obtained, an integrator is incorporated into the filter structure. To compute the filter gains, SOS techniques have been used to reduce the problems to SDP. The effectiveness of the proposed method is confirmed through simulation examples.

5.2 SYSTEM DESCRIPTION AND PROBLEM FORMULATION

Consider the following polynomial discrete-time system:

$$\begin{aligned} x(k+1) &= A(x(k))x(k) + B(x(k))w(k), \\ z &= C_1(x(k))x(k), \qquad (5.1) \\ y &= C(x(k))x(k) + D(x(k))w(k), \qquad (5.2) \end{aligned}$$

where $x(k) \in \Re^n$ is the state, $z(k)$ is the regulated output, $w(k)$ is the disturbance, y is the measurement, and $A(x(k))$, $B(x(k))$, $C_1(x(k))x(k)$, $C(x(k))$, and $D(x(k))$ are polynomial matrices of appropriate dimensions.

The dynamic filter is given as

$$\begin{aligned} \hat{x}(k+1) &= A(\hat{x}(k))\hat{x} + L(\hat{x}(k))\big(y - \hat{y}\big), \\ \hat{y} &= C(\hat{x}(k))\hat{x}(k), \qquad (5.3) \end{aligned}$$

where $\hat{x}$ is a filter state, $\hat{y}$ is a filter measurement, and $L(\hat{x}(k))$ is a designed polynomial matrix of appropriate dimensions.

H_∞ Problem Formulation. Given a prescribed performance $\gamma > 0$, design a dynamic filter of the form (5.3) such that

$$\|z(k)\|_{[0,\infty]} \leq \gamma^2 \|\omega(k)\|_{[0,\infty]} . \qquad (5.4)$$

The dynamics of the estimation error defined by $e = \hat{x}(k) - x(k)$ is given as

$$\begin{aligned} e(k+1) &= \hat{x}(k+1) - x(k+1) \\ &= A(\hat{x}) + L(\hat{x})\big(C(\hat{x})\hat{x} - C(x)x\big) - A(x)x - B(x)w \end{aligned}$$

$$= \big[A(\hat{x}) + L(\hat{x})C(\hat{x})\big]e + \big[A(\hat{x}) - A(x) + L(\hat{x})C(\hat{x}) - L(\hat{x})C(x)\big]x - L(x)D(x)w - B(x)w. \tag{5.5}$$

Now, letting $\tilde{e} = [e^T, x^T]^T$, system (5.5) can be rewritten as

$$\tilde{e}(k+1) = \phi(x, \hat{x})e_w, \tag{5.6}$$

where

$$\phi(x, \hat{x}) = \begin{bmatrix} A(\hat{x}) + L(\hat{x})C(\hat{x}) & \begin{array}{c} A(\hat{x}) - A(x) \\ +L(\hat{x})[C(\hat{x}) - C(x)] \end{array} & -L(\hat{x})D(x) - B(x) \\ 0 & A(x) & B(x) \end{bmatrix} \tag{5.7}$$

and $e_w = [e^T \;\; x^T \;\; w^T]^T$.

Theorem 12. *Given $\gamma > 0$, the estimation error dynamic (5.6) is asymptotically stable with the prescribed H_∞ performance γ if there exist polynomial matrices $L(\hat{x})$ and $P(\tilde{e})$ such that the following conditions hold:*

$$P(\tilde{e}) > 0, \tag{5.8}$$

$$\begin{bmatrix} \tilde{P}(\tilde{e}) & \phi^T(x, \hat{x})P(\tilde{e}(k+1)) \\ P(\tilde{e}(k+1))\phi(x, \hat{x}) & P(\tilde{e}(k+1)) \end{bmatrix} > 0, \tag{5.9}$$

where

$$\tilde{P}(\tilde{e}) = \begin{bmatrix} P(\tilde{e}) & 0 \\ 0 & \gamma^2 I \end{bmatrix} - \bar{C}_1^T(x, \hat{x})\bar{C}_1(x, \hat{x}) \tag{5.10}$$

and

$$\bar{C}_1(x, \hat{x}) = \big[C_1(\hat{x}) \;\; [C_1(\hat{x}) - C_1(x)] \;\; 0\big]. \tag{5.11}$$

Proof. Choose the Lyapunov function as

$$V(\tilde{e}) = \tilde{e}^T P(\tilde{e})\tilde{e}. \tag{5.12}$$

The time difference of $V(\tilde{e}(k))$ along (5.6) is given as

$$\begin{aligned} \Delta(V(\tilde{e})) &= V(\tilde{e}(k+1)) - V(\tilde{e}) \\ &= \tilde{e}^T(k+1)P(\tilde{e}(k+1))\tilde{e}(k+1) - \tilde{e}^T P(\tilde{e})\tilde{e} \\ &= e_w^T\phi^T(x, \hat{x})P(\tilde{e}(k+1))\phi(x, \hat{x})e_w - \tilde{e}^T P(\tilde{e})\tilde{e}. \end{aligned} \tag{5.13}$$

Adding and subtracting $z^{(}k)z(k) + \gamma^2 w^T(k)w(k)$ to and from (5.13) yield

$$\Delta(V(\tilde{e})) = e_w^T\big[\phi^T(x,\hat{x})P(\tilde{e}(k+1))\phi(x,\hat{x}) - \tilde{P}(\tilde{e})\big]e_w \\ - z^T(k)z(k) + \gamma^2 w^T(k)w(k). \tag{5.14}$$

Suppose (5.9) is feasible. Then applying the Schur complement, we have

$$\phi^T(x,\hat{x})P(\tilde{e}(k+1))\phi(x,\hat{x}) - P(\tilde{e}) < 0. \tag{5.15}$$

Knowing that (5.15) holds, we have

$$\Delta(V(\tilde{e})) \leq -z^T(k)z(k) + \gamma^2 w^T(k)w(k). \tag{5.16}$$

Taking the summation on both sides of (5.16) and assuming that $\tilde{e}(0) = 0$, we have

$$\|z(k)\|_{[0,\infty]} \leq \gamma^2 \|\omega(k)\|_{[0,\infty]}. \tag{5.17}$$

This concludes that the estimation error dynamics meets the prescribed H_∞ performance. When $w = 0$, it is easy to see that $\Delta(V(\tilde{e})) < 0$, which implies that the estimation error dynamic is asymptotically stable. This completes the proof. □

Theorem 12 is hard to solve because of the product of $P(\tilde{e})$ and $L(\hat{x})$. To solve this problem, an integrator is incorporated into the filter dynamics. In doing so, a convex solution to the filter design problem for polynomial discrete-time systems can be rendered efficiently. The details of this integrator method are illustrated in the following section.

5.3 MAIN RESULTS

In this section, we illustrate the significance of incorporating an integrator into the H_∞ filter dynamics for polynomial discrete-time systems with and without uncertainties.

5.3.1 Nonlinear H_∞ filtering for polynomial discrete-time systems

We propose following nonlinear filter dynamic with an integrator:

$$\hat{x}(k+1) = A(\hat{x})\hat{x} + x_f, \\ x_f(k+1) = x_f + L(\hat{x})\big(C(\hat{x})\hat{x} - y\big), \tag{5.18}$$

where $\hat{x}$ is a filter state, x_f is an augmented integrator state, $L(\hat{x})$ is a designed polynomial matrix, and $A(\hat{x})$, $C(\hat{x})$, and $C(x)$ are all polynomial matrices of appropriate dimensions.

Now, the estimation error dynamics is defined as follows:

$$\bar{e} = \begin{bmatrix} e_1 \\ e_2 \end{bmatrix} = \begin{bmatrix} \hat{x} - x \\ x_f \end{bmatrix}. \tag{5.19}$$

The error dynamics is then given by

$$\begin{aligned} \bar{x}(k+1) &= \begin{bmatrix} e_1(k+1) \\ e_2(k+1) \end{bmatrix} = \begin{bmatrix} \hat{x}(k+1) - x(k+1) \\ x_f(k+1) \end{bmatrix} \\ &= \begin{bmatrix} A(\hat{x})\hat{x} + x_f - A(x)x - B(x)w \\ x_f + L(\hat{x})\big(C(\hat{x})\hat{x} - C(x)x\big) \end{bmatrix} \\ &= \begin{bmatrix} A(\hat{x})e_1 + e_2 + \big(A(\hat{x}) - A(x)\big)x - B(x)w \\ e_2 + L(\hat{x})C(\hat{x})e_1 + \big(L(\hat{x})C(\hat{x}) - L(\hat{x})C(x)\big)x - L(\hat{x})D(x)w \end{bmatrix}. \end{aligned} \tag{5.20}$$

Next, defining $\check{e} = [e_1^T x^T e_2^T]^T$, we can rewrite the estimation error dynamics described in (5.20) as follows:

$$\check{e}(k+1) = [\phi_1(x, \hat{x}) + \phi_2(x, \hat{x})]e_w, \tag{5.21}$$

where, $e_w = [\check{e}^T w^T]^T$,

$$\phi_1(x, \hat{x}) = \begin{bmatrix} 0 & 0 & 0 & 0 \\ 0 & 0 & 0 & 0 \\ L(\hat{x})C(\hat{x}) & L(\hat{x})C(\hat{x}) - L(\hat{x})C(x) & 0 & -L(\hat{x})D(x) \end{bmatrix}, \tag{5.22}$$

and

$$\phi_2(x, \hat{x}) = \begin{bmatrix} A(\hat{x}) & A(\hat{x}) - A(x) & I & -B(x) \\ 0 & A(x) & 0 & B(x) \\ 0 & 0 & I & 0 \end{bmatrix}. \tag{5.23}$$

Theorem 13. *Given $\gamma > 0$, the estimation error dynamics (5.21) are asymptotically stable with the prescribed H_∞ performance γ if there exist symmetric polynomial matrix $P(e_1)$ and polynomial matrices $K(\hat{x})$ and $G(\hat{x})$ such that the following conditions hold:*

$$P(e_1) > 0, \tag{5.24}$$

$$\begin{bmatrix} \tilde{P}(e_1) & \tilde{\phi}_1^T(x, \hat{x}) + \phi_2^T(x, \hat{x})G^T(\hat{x}) \\ \tilde{\phi}_1(x, \hat{x}) + G(\hat{x})\phi_2(x, \hat{x}) & G^T(\hat{x}) + G(\hat{x}) - P(e_1(k+1)) \end{bmatrix} > 0, \tag{5.25}$$

where

$$G(\hat{x}) = \begin{bmatrix} G_{11}(\hat{x}) & G_{12}(\hat{x}) & G_{13}(\hat{x}) \\ G_{21}(\hat{x}) & G_{22}(\hat{x}) & G_{13}(\hat{x}) \\ G_{31}(\hat{x}) & G_{32}(\hat{x}) & G_{13}(\hat{x}) \end{bmatrix} \tag{5.26}$$

and

$$\tilde{\phi}_1(x,\hat{x}) = \begin{bmatrix} K(\hat{x})C(\hat{x}) & K(\hat{x})[C(\hat{x}) - C(x)] & 0 & 0 \\ K(\hat{x})C(\hat{x}) & K(\hat{x})[C(\hat{x}) - C(x)] & 0 & 0 \\ K(\hat{x})C(\hat{x}) & K(\hat{x})[C(\hat{x}) - C(x)] & 0 & K(\hat{x})D(x) \end{bmatrix}. \tag{5.27}$$

Moreover, the dynamic filter is given by

$$\begin{aligned} \hat{x}(k+1) &= A(\hat{x})\hat{x} + x_f, \\ x_f(k+1) &= x_f + L(\hat{x})\big(C(\hat{x})\hat{x} - y\big), \end{aligned} \tag{5.28}$$

where

$$L(\hat{x}) = K(\hat{x})G_{13}^{-1}(\hat{x}). \tag{5.29}$$

Proof. Select a Lyapunov function as

$$V(\check{e}) = \check{e}^T P(e_1)\check{e}. \tag{5.30}$$

Then the time difference of $V(\check{e}(k))$ along (5.21) with (5.18) is

$$\begin{aligned} \Delta(V(\check{e})) &= V(\check{e}(k+1)) - V(\check{e}) \\ &= \check{e}^T(k+1)P(e_1(k+1))\check{e}(k+1) - \check{e}^T P(e_1)\check{e} \\ &= e_w^T(\phi_1(x,\hat{x}) + \phi_2(x,\hat{x}))^T P(e_1(k+1))(\phi_1(x,\hat{x}) + \phi_2(x,\hat{x}))e_w \\ &- e_w^T P(\check{e})e_w. \end{aligned} \tag{5.31}$$

Adding and subtracting $z^(k)z(k) + \gamma^2 w^T(k)w(k)$ to and form (5.18) yield

$$\begin{aligned} \Delta(V(\tilde{e})) &= e_w^T(\phi_1(x,\hat{x}) + \phi_2(x,\hat{x}))^T P(e_1(k+1))(\phi_1(x,\hat{x}) + \phi_2(x,\hat{x}))e_w \\ &- e_w^T \tilde{P}(\check{e})e_w - z^T(k)z(k) + \gamma^2 w^T(k)w(k). \end{aligned} \tag{5.32}$$

Suppose (5.25) is feasible; thus, $G^T(\hat{x}) + G(\hat{x}) > P(e_1(k+1)) > 0$. This implies that $G(\hat{x})$ is nonsingular. Since $P(e_1(k+1))$ is positive definite, we have the inequality

$$\big(P(e_1(k+1)) - G(\hat{x})\big)P^{-1}(e_1(k+1))\big(P(e_1(k+1)) - G(\hat{x})\big)^T > 0 \tag{5.33}$$

or

$$G(\hat{x}(k))P^{-1}(e_1(k+1))G^T(\hat{x}) \geq G(\hat{x}) + G^T(\hat{x}) - P(e_1(k+1)). \quad (5.34)$$

This immediately gives

$$\begin{bmatrix} \tilde{P}(e_1) & \tilde{\phi}_1^T(x,\hat{x}) + \phi_2^T(x,\hat{x})G^T(\hat{x}) \\ \tilde{\phi}_1(x,\hat{x}) + G(\hat{x})\phi_2(x,\hat{x}) & G^T(\hat{x})P^{-1}(e_1(k+1))G(\hat{x}) \end{bmatrix} > 0. \quad (5.35)$$

Next, multiplying (5.35) on the right by $diag[I, G^{-1}(\hat{x}(k))]^T$ and on the left by $diag[I, G^{-1}(\hat{x}(k))]$, we get

$$\begin{bmatrix} \tilde{P}(e_1) & \phi_1^T(x,\hat{x}) + \phi_2^T(x,\hat{x}) \\ \phi_1(x,\hat{x}) + \phi_2(x,\hat{x}) & P^{-1}(e_1(k+1)) \end{bmatrix} > 0. \quad (5.36)$$

Then, applying the Schur complement in (5.36), we have

$$(\phi_1(x,\hat{x}) + \phi_2(x,\hat{x}))^T P(e_1(k+1))(\phi_1(x,\hat{x}) + \phi_2(x,\hat{x})) - \tilde{P}(e_1) < 0. \quad (5.37)$$

Knowing that (5.37) holds, we have

$$\Delta(V(\tilde{e})) \leq -z^T(k)z(k) + \gamma^2 w^T(k)w(k). \quad (5.38)$$

Taking the summation on both sides of (5.38) and assuming that $\tilde{e}(0) = 0$, we have

$$\|z(k)\|_{[0.\infty]} \leq \gamma^2 \|\omega(k)\|_{[0.\infty]} \cdot \quad (5.39)$$

This concludes that the estimation error dynamics meets the prescribed H_∞ performance. When $w = 0$, it is easy to see that $\Delta(V(\tilde{e})) < 0$, which implies that the estimation error dynamic is asymptotically stable. This completes the proof. □

Inequalities (5.25) of Theorem 13 are convex. For clarity, refer to the following expansion version of $P(e_1(k+1))$:

$$P(e_1(k+1)) = P\big[A(\hat{x})\hat{x} + x_f - A(x)x\big]. \quad (5.40)$$

From (5.40), x_f is an augmented state; hence, $P(e_1(k+1))$ provides a convex solution, so that Theorem 13 can be possibly solved via SDP. To relax these conditions, we utilize the SOS decomposition approach [30], and the conditions given in Theorem 13 can be converted into SOS conditions, which are given by the following corollary.

Corollary 6. *Given $\gamma > 0$, the estimation error dynamics (5.21) is asymptotically stable with the prescribed H_∞ performance γ if there exist asymmetric polynomial matrix $P(e_1)$, polynomial matrices $K(\hat{x})$ and $G(\hat{x})$, and positive constants ϵ_1 and ϵ_2 such that the following conditions are SOSs:*

$$v_1^T[P(e_1) - \epsilon_1 I]v_1, \tag{5.41}$$

$$v_2^T \begin{bmatrix} \tilde{P}(e_1) - \epsilon_2 I & \tilde{\phi}_1^T(x, \hat{x}) + \phi_2^T(x, \hat{x})G^T(\hat{x}) \\ \tilde{\phi}_1(x, \hat{x}) + G(\hat{x})\phi_2(x, \hat{x}) & G^T(\hat{x}) + G(\hat{x}) - P(e_1(k+1)) - \epsilon_2 I \end{bmatrix} v_2, \tag{5.42}$$

where, v_1 and v_2 are free vectors of appropriate dimensions. Furthermore, the filter is given by

$$\begin{aligned} \hat{x}(k+1) &= A(\hat{x})\hat{x} + x_f, \\ x_f(k+1) &= x_f + L(\hat{x})\big(C(\hat{x})\hat{x} - y\big), \end{aligned} \tag{5.43}$$

where

$$L(\hat{x}) = K(\hat{x})G_{12}^{-1}(\hat{x}). \tag{5.44}$$

In the next section, we consider a polynomial discrete-time system with uncertainties that satisfy integral functional constraints.

5.3.2 Robust nonlinear H_∞ filtering for polynomial discrete-time systems

Consider the following polynomial discrete-time system with uncertainties that satisfy integral functional constraints:

$$\begin{aligned} x(k+1) &= A(x(k))x(k) + \Delta A(x(k))x(k) + B(x)w(k), \\ z &= C_1(x(k))x(k), \\ y &= C(x(k))x(k) + \Delta C(x(k), \theta)x(k) + D(x(k))w(k), \end{aligned} \tag{5.45}$$

where $\Delta A(x(k))$ and $\Delta Cx(k))$ are

$$\begin{bmatrix} \Delta A(x) \\ \Delta C(x) \end{bmatrix} = \begin{bmatrix} H_1(x) \\ H_2(x) \end{bmatrix} F(x, t)E(x) \tag{5.46}$$

with known matrix functions $H_1(x)$, $H_2(x)$, and $E(x)$ that characterize the structure of the uncertainties and $E(0) = 0$. Further, the following integral functional constraint holds:

$$\|E(x(k))\|_{[0,\infty]} - \|F(x(k), k)E(x(k))\|_{[0,\infty]} \geq 0. \tag{5.47}$$

In [31], the authors have established the interconnection between the robust nonlinear H_∞ filtering problem and the nonlinear H_∞ filtering problem for known systems, i.e., systems without uncertainties. Motivated by this result, we define the scaled system

$$\begin{aligned} x(k+1) &= A(x(k))x(k) + \tilde{B}(x)w(k), \\ z &= \tilde{C}_1(x(k))x(k), \\ y &= C(x(k))x(k) + \tilde{D}(x(k))w(k), \end{aligned} \tag{5.48}$$

where $\tilde{B}(x) = [B(x)\frac{1}{\delta\gamma}H_1(x)]$, $\tilde{D}(x) = [D(x) \;\; \frac{1}{\delta\gamma}H_2(x)]$, and $\tilde{C}_1(x) = \begin{bmatrix} C_1(x(k)) \\ \delta E(x(k)) \end{bmatrix}$ with $\delta > 0$. Based on [31], the robust H_∞ filtering problem for (5.45) is solvable if and only if the H_∞ filtering problem for (5.48) is solvable. Hence, we have the following theorem.

Theorem 14. *Given $\gamma > 0$ and $\delta > 0$, the robust filtering problem for the system (5.45) is solvable if there exist symmetric polynomial matrices $P(e_1)$ and polynomial matrices $K(\hat{x})$ and $G(\hat{x})$ such that the following conditions hold:*

$$P(e_1) > 0, \tag{5.49}$$

$$\begin{bmatrix} \tilde{P}(e_1) & \tilde{\phi}_1^T(x,\hat{x}) + \phi_2^T(x,\hat{x})G^T(\hat{x}) \\ \tilde{\phi}_1(x,\hat{x}) + G(\hat{x})\phi_2(x,\hat{x}) & G^T(\hat{x}) + G(\hat{x}) - P(e_1(k+1)) \end{bmatrix} > 0, \tag{5.50}$$

where

$$G(\hat{x}) = \begin{bmatrix} G_{11}(\hat{x}) & G_{12}(\hat{x}) & G_{13}(\hat{x}) \\ G_{21}(\hat{x}) & G_{22}(\hat{x}) & G_{13}(\hat{x}) \\ G_{31}(\hat{x}) & G_{32}(\hat{x}) & G_{13}(\hat{x}) \end{bmatrix}, \tag{5.51}$$

$$\phi_2(x,\hat{x}) = \begin{bmatrix} A(\hat{x}) & A(\hat{x}) - A(x) & I & -\tilde{B}(x) \\ 0 & A(x) & 0 & \tilde{B}(x) \\ 0 & 0 & I & 0 \end{bmatrix}, \tag{5.52}$$

$$\tilde{\phi}_1(x,\hat{x}) = \begin{bmatrix} K(\hat{x})C(\hat{x}) & K(\hat{x})[C(\hat{x}) - C(x)] & 0 & 0 \\ K(\hat{x})C(\hat{x}) & K(\hat{x})[C(\hat{x}) - C(x)] & 0 & 0 \\ K(\hat{x})C(\hat{x}) & K(\hat{x})[C(\hat{x}) - C(x)] & 0 & K(\hat{x})\tilde{D}(x) \end{bmatrix}, \tag{5.53}$$

$$\tilde{P}(\tilde{e}) = \begin{bmatrix} P(\tilde{e}) & 0 \\ 0 & \gamma^2 I \end{bmatrix} - \tilde{C}_1^T(x,\hat{x})\tilde{C}_1(x,\hat{x}), \tag{5.54}$$

and

$$\bar{C}_1(x,\hat{x}) = \left[\tilde{C}_1(\hat{x})\ \ [\tilde{C}_1(\hat{x}) - \tilde{C}_1(x)]\ \ 0\right]. \tag{5.55}$$

Moreover, the dynamic filter is given by

$$\begin{aligned}\hat{x}(k+1) &= A(\hat{x})\hat{x} + x_f,\\ x_f(k+1) &= x_f + L(\hat{x})\big(C(\hat{x})\hat{x} - y\big),\end{aligned} \tag{5.56}$$

where

$$L(\hat{x}) = K(\hat{x})G_{13}^{-1}(\hat{x}). \tag{5.57}$$

Similarly, Theorem 14 can be converted into SOS conditions, which are given in the following corollary.

Corollary 7. *Given $\gamma > 0$ and $\delta > 0$, the robust filtering problem for the system (5.45) is solvable if there exist asymmetric polynomial matrix $P(e_1)$, polynomial matrices $K(\hat{x})$ and $G(\hat{x})$, and positive constants ϵ_1 and ϵ_2 such that the following conditions are SOSs:*

$$v_1^T[P(e_1) - \epsilon_1 I]v_1, \tag{5.58}$$

$$v_2^T\begin{bmatrix} \tilde{P}(e_1) - \epsilon_2 I & \tilde{\phi}_1^T(x,\hat{x}) + \phi_2^T(x,\hat{x})G^T(\hat{x}) \\ \tilde{\phi}_1(x,\hat{x}) + G(\hat{x})\phi_2(x,\hat{x}) & G^T(\hat{x}) + G(\hat{x}) - P(e_1(k+1)) - \epsilon_2 I \end{bmatrix} v_2, \tag{5.59}$$

where v_1 and v_2 are free vectors of appropriate dimensions. Furthermore, the filter is given by

$$\begin{aligned}\hat{x}(k+1) &= A(\hat{x})\hat{x} + x_f,\\ x_f(k+1) &= x_f + L(\hat{x})\big(C(\hat{x})\hat{x} - y\big),\end{aligned} \tag{5.60}$$

where

$$L(\hat{x}) = K(\hat{x})G_{12}^{-1}(\hat{x}). \tag{5.61}$$

5.4 NUMERICAL EXAMPLES

Example 1. Consider a tunnel diode circuit shown in Fig. 5.1, where the characteristics of the tunnel diode are described as follows:

$$i_D(t) = 0.002v_D(t) + 0.01v_D^3(t). \tag{5.62}$$

Next, letting $x_1(t)$=$v_c(t)$ and $x_2(t)$=$i_L(t)$ be the state variables, the circuit

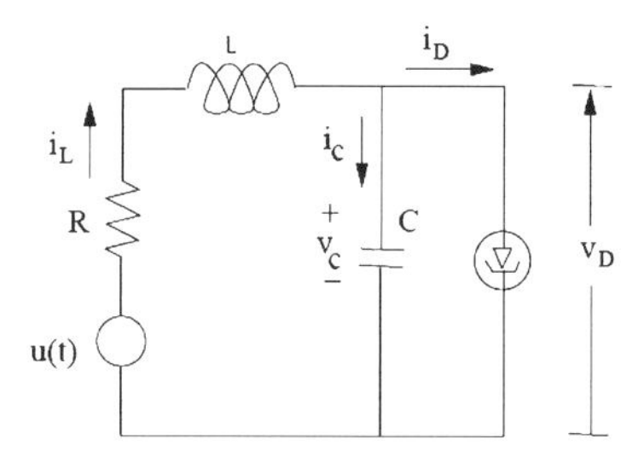

■ **FIGURE 5.1** A tunnel diode circuit.

is governed by the following state equations:

$$
\begin{aligned}
C\dot{x}_1(t) &= -0.002x_1(t) - 0.01x_1^3(t) + x_2(t),\\
L\dot{x}_2(t) &= -x_1(t) - Rx_2(t),\\
z(t) &= x_2(t),\\
y(t) &= x_2(t),
\end{aligned}
\tag{5.63}
$$

where $z(t)$ is the regulated output, $y(t)$ is the measurement, $x_1(t)=v_c(t)$, and $x_2(t)=i_L(t)$. Meanwhile the circuit parameter is given as follows: $C =$ 20 mF, $L = 1000$ mH, and $R = 1\ \Omega$. With these parameters, the dynamics of the circuit can be written as follows:

$$
\begin{aligned}
\dot{x}_1(t) &= -0.1x_1(t) - 0.5x_1^3(t) + 50x_2(t) + \omega_1(t),\\
\dot{x}_2(t) &= -x_1(t) - x_2(t) + \omega_1(t),\\
z(t) &= x_2(t),\\
y(t) &= x_2(t) + \omega_2(t),
\end{aligned}
\tag{5.64}
$$

where $\omega_(t)$ and $\omega_2(t)$ are the noises. By Euler's discretization method, the above system is sampled at $T = 0.02$, that is,

$$
\begin{aligned}
x_1(k+1) &= x_1(k) + T\big[-0.1x_1(k) - 0.5x_1^3(k) + 50x_2(k) + \omega_1(k)\big],\\
x_2(k+1) &= x_2(k) + T\big[-x_1(k) - x_2(k) + \omega_1(k)\big],\\
z(k) &= x_2(k),\\
y(k) &= x_2(k) + \omega_2(k).
\end{aligned}
\tag{5.65}
$$

Applying the procedures outlined in Corollary 6 with $\gamma = 1$ and $\epsilon_1 = \epsilon_2 =$ 0.01, $P(e_1(k))$ and $G(\hat{x}(k))$ are set to be of degree 5, and $K(\hat{x}(k))$ is selected to be of degree 10. Through this setup, a feasible solution is obtained. The solutions are to large to be included here. The ratio of the estimation error energy to the noise energy is shown in Fig. 5.2. It can be clearly seen

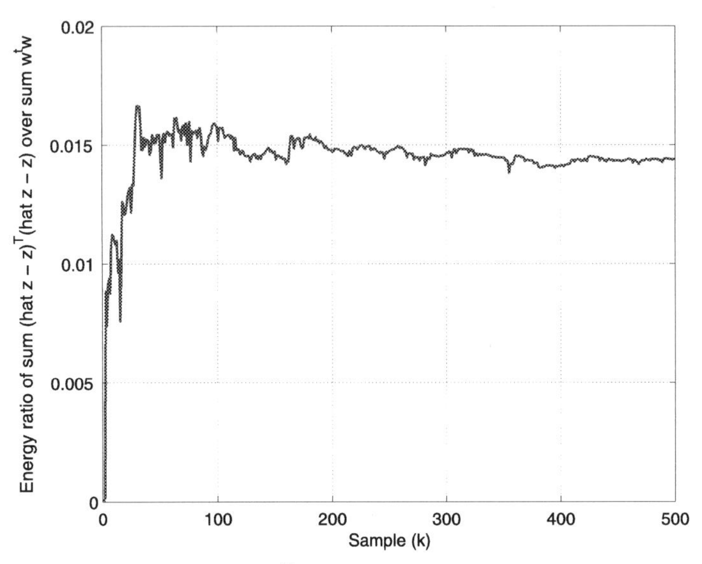

■ **FIGURE 5.2** Energy ratio: $\frac{\sum(z-\hat{z})^T(z-\hat{z})}{\sum \omega^T\omega}$.

from the figure that the energy ratio tends to be a constant value after 400 samples, which is approximately 0.0142. Hence, the γ value is equivalent to $\sqrt{0.0142} \approx 0.1192$, which is less than the prescribed γ value 1.

Example 2. Consider the same tunnel diode circuit as in Example 1. Here we assume that $R = 1 \pm 30\%\ \Omega$. The dynamics of the circuit can be written as

$$\begin{aligned} \dot{x}_1(t) &= -0.1x_1(t) - 0.5x_1^3(t) + 50x_2(t) + \omega_1(t), \\ \dot{x}_2(t) &= -x_1(t) - (1+\Delta R)x_2(t) + \omega_1(t), \\ z(t) &= x_2(t), \\ y(t) &= x_2(t) + \omega_2(t). \end{aligned} \tag{5.66}$$

Similarly, the above system is sampled at $T = 0.02$ by Euler's discretization method as

$$\begin{aligned} x_1(k+1) &= x_1(k) + T\big[-0.1x_1(k) - 0.5x_1^3(k) + 50x_2(k) + \omega_1(k)\big], \\ x_2(k+1) &= x_2(k) + T\big[-x_1(k) - (1+\Delta R)x_2(k) + \omega_1(k)\big], \\ z(k) &= x_2(k), \\ y(k) &= x_2(k) + \omega_2(k). \end{aligned} \tag{5.67}$$

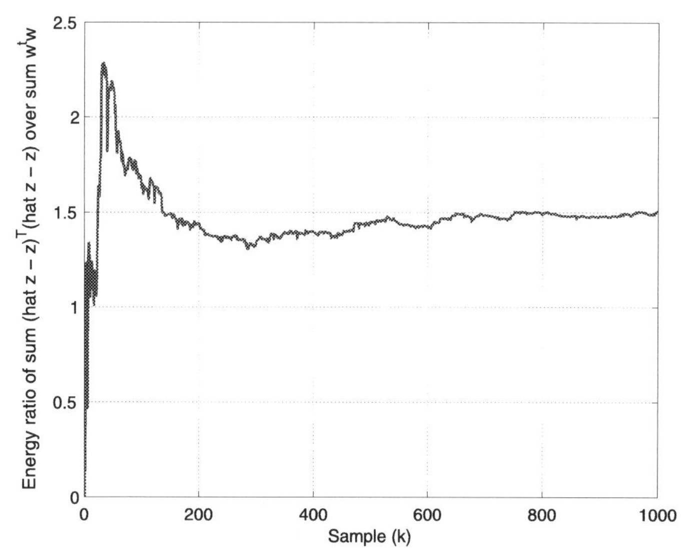

■ **FIGURE 5.3** Robust energy ratio: $\frac{\sum (z-\hat{z})^T (z-\hat{z})}{\sum \omega^T \omega}$.

Express ΔR in the form of (5.46) with

$$H_1(x) = \begin{bmatrix} 0 \\ 0.3T \end{bmatrix}, \quad E(x(k)) = \begin{bmatrix} 0 & 1 \end{bmatrix}, \quad H_2(x) = 0.$$

Based on [31], we define

$$\begin{aligned} \bar{B}(x) &= \begin{bmatrix} B(x) & \frac{1}{\delta\gamma} H_1(x) \end{bmatrix}, \quad \bar{C}(x) = \begin{bmatrix} C_1(x) \\ \delta E(x) \end{bmatrix}, \\ \bar{D}(x) &= \begin{bmatrix} D(x) & \frac{1}{\delta\gamma} H_2(x) \end{bmatrix}. \end{aligned} \tag{5.68}$$

Applying Corollary 7 with $\delta = 1$, $\gamma = 2$, and $\epsilon_1 = \epsilon_2 = 0.01$, $P(e_1(k))$ and $G(\hat{x}(k))$ are set to be of degree of 6, and $K(\hat{x}(k))$ is selected to be of degree 12. Fig. 5.3 shows the ratio of the estimation error energy to the noise energy. The energy ratio tends to 1.5 after 100 samples, that is, $\gamma = \sqrt{1.5} \approx 1.2247$, which is less than the prescribed γ value 2.

5.5 CONCLUSION

The nonlinear H_∞ filtering design problem for polynomial nonlinear systems has been examined in this chapter. We also consider polynomial nonlinear systems with uncertainties that are described by some integral functional constraints. Sufficient conditions for the existence of a robust H_∞ nonlinear filter are given in terms of the solvability of the PMIs, which is formulated as SOS constraints that can be solved by any SOS solver. Convex solutions to the nonlinear H_∞ filtering design problem can be solved efficiently by incorporating an integrator into the filter structure.

REFERENCES

[1] J.A. Ball, J.W. Helton, H_∞ control for nonlinear plants: connection with differential games, in: Proc. 28th IEEE Conf. Decision Control, Tampa, FL, 1989, pp. 956–962.

[2] T. Basar, G.J. Olsder, Dynamic Noncooperative Game Theory, Academic Press, New York, 1982.

[3] A.J. van der Schaft, L_2-gain analysis of nonlinear systems and nonlinear state feedback H_∞ control, IEEE Transactions on Automatic Control 31 (1992) 770–784.

[4] A. Isidori, A. Astolfi, Disturbance attenuation and H_∞-control via measurement feedback in nonlinear systems, IEEE Transactions on Automatic Control 31 (1992) 1283–1293.

[5] A. Isidori, Feedback control of nonlinear systems, in: Proc. First European Control Conf., Grenoble, France, 1991, pp. 1001–1012.

[6] T. Basar, P. Bernhard, H_∞ optimal control and related minimax design problems, in: Systems and Control: Foundations and Applications, Birkhauser, Boston, 1991.

[7] J.C. Willems, Dissipative dynamical systems. Part I: general theory, Archive for Rational Mechanics and Analysis 45 (1992) 321–351.

[8] D.J. Hill, P.J. Moylan, Dissipative dynamical systems: basic input–output and state properties, Journal of the Franklin Institute 309 (1980) 327–357.

[9] J.A. Ball, J.W. Helton, M.L. Walker, H_∞ control for nonlinear systems with output feedback, IEEE Transactions on Automatic Control 38 (1993) 546–559.

[10] F. Rasool, D. Huang, S.K. Nguang, Robust H_∞ output feedback control of networked control systems with multiple quantizers, Journal of the Franklin Institute 349 (3) (2012) 1153–1173.

[11] S.K. Nguang, W. Assawinchaichote, P. Shi, Y. Shi, Robust H_∞ control design for uncertain fuzzy systems with Markovian jumps: an LMI approach, in: American Control Conference, 2005, pp. 1805–1810.

[12] S.K. Nguang, P. Zhang, S.X. Ding, Parity relation based fault estimation for nonlinear systems: an LMI approach, International Journal of Automation and Computing 4 (2) (2007) 164–168.

[13] S.K. Nguang, Comments on "Robust stabilization of uncertain input-delay systems by sliding mode control with delay compensation", Automatica 37 (10) (2001) 1677.

[14] S.K. Nguang, P. Shi, H_∞ output feedback control of fuzzy system models under sampled measurements, Computers and Mathematics With Applications 46 (5) (2003) 705–717.

[15] W. Assawinchaichote, S.K. Nguang, P. Shi, Fuzzy Control and Filter Design for Uncertain Fuzzy Systems, Springer, 2006.

[16] S.K. Nguang, P. Shi, On designing filters for uncertain sampled-data nonlinear systems, Systems & Control Letters 41 (5) (2000) 305–316.

[17] D. Huang, S.K. Nguang, Robust Control for Uncertain Networked Control Systems With Random Delays, Springer Science & Business Media, 2009.

[18] J. Zhang, A.K. Swain, S.K. Nguang, Robust sensor fault estimation scheme for satellite attitude control systems, Journal of the Franklin Institute 350 (9) (2013) 2581–2604.

[19] J. Zhang, P. Shi, J. Qiu, S.K. Nguang, A novel observer-based output feedback controller design for discrete-time fuzzy systems, IEEE Transactions on Fuzzy Systems 23 (1) (2015) 223–229.

[20] S.K. Nguang, P. Shi, Delay-dependent H_∞ filtering for uncertain time delay nonlinear systems: an LMI approach, IET Control Theory & Applications 1 (1) (2007) 133–140.

[21] Z. Hou, J. Luo, P. Shi, S.K. Nguang, Stochastic stability of Ito differential equations with semi-Markovian jump parameters, IEEE Transactions on Automatic Control 51 (8) (2006) 1383–1387.

[22] W. Assawinchaichote, S.K. Nguang, P. Shi, E.K. Boukas, H_∞ fuzzy state-feedback control design for nonlinear systems with stability constraints: an LMI approach, Mathematics and Computers in Simulation 78 (4) (2008) 514–531.

[23] S. Chae, S.K. Nguang, SOS based robust H_∞ fuzzy dynamic output feedback control of nonlinear networked control systems, IEEE Transactions on Cybernetics 44 (7) (2014) 1204–1213.

[24] F. Rasool, D. Huang, S.K. Nguang, Robust H_∞ output feedback control of discrete-time networked systems with limited information, Systems & Control Letters 60 (10) (2011) 845–853.

[25] S.K. Nguang, P. Shi, Stabilisation of a class of nonlinear time-delay systems using fuzzy models, in: Proceedings of the 39th IEEE Conference on Decision and Control, 2000, pp. 5–11.

[26] S. Saat, S.K. Nguang, Nonlinear H_∞ output feedback control with integrator for polynomial discrete-time systems, International Journal of Robust and Nonlinear Control 25 (2015) 1051–1065.

[27] Y. Zhang, P. Shi, S.K. Nguang, H.R. Karimi, Observer-based finite-time fuzzy H_∞ control for discrete-time systems with stochastic jumps and time-delays, Signal Processing 97 (2014) 252–261.

[28] S. Chae, F. Rasool, S.K. Nguang, A. Swain, Robust mode delay-dependent H_∞ control of discrete-time systems with random communication delays, IET Control Theory & Applications 4 (6) (2010) 936–944.

[29] Y. Zhang, P. Shi, S.K. Nguang, Y. Song, Robust finite-time H_∞ control for uncertain discrete-time singular systems with Markovian jumps, IET Control Theory & Applications 8 (12) (2014) 1105–1111.

[30] P.A. Parrilo, Structured Semidefinite Programs and Semialgebraic Geometry Methods in Robustness and Optimization, PhD dissertation, California Inst. Technol., Pasadena, 2000.

[31] S.K. Nguang, M. Fu, Robust nonlinear H_∞ filtering, Automatica 32 (8) (1996) 1195–1199.

Chapter 6

Robust nonlinear H_∞ output feedback control for polynomial discrete-time systems

CHAPTER OUTLINE

6.1 INTRODUCTION

In many practical systems, states are not all available for feedback. Due to this reason, the static output feedback design has attracted much attention from the control practitioners. A comprehensive survey on static output feedback can be found in [1]. In [1] the authors prove that any dynamic output feedback problem can be transformed into a static output feedback problem. Hence, the static output feedback formulation provides a more general method than the full-order dynamic output feedback [2–21], and therefore the static output formulation can be applied to design a full-order dynamic controller. The converse, however, is not true.

Although the static output feedback control design for polynomial systems is not as widely studied as its linear counterpart, some significant attempts can still be found in [22–24]. In particular, to achieve a convex solution

Analysis and Synthesis of Polynomial Discrete-Time Systems. DOI: 10.1016/B978-0-08-101901-6.00006-2

to the H_∞ control problem, a predefined upper bound has been introduced in [22] to limit the effect of the nonconvex term. However, this predefined upper bound is hard to determine beforehand, and the closed-loop stability can only be guaranteed within a bound region. On the other hand, in [23,24] the existence of a nonlinear static output feedback control law is given by the solvability of polynomial matrix inequalities (PMIs) formulated as SOS constraints. To solve the nonconvexity that exists in the PMIs, an iterative algorithm based on the SOS decomposition has been developed. Unfortunately, it is hard to determine the first value of a slack variable matrix $\epsilon(x)$ because it is unknown. The value of $\epsilon(x)$ plays a vital role in this approach because it determines the feasibility of the problem. There is no unique way to choose this value, and thus it is difficult to apply this approach.

The above-mentioned results are concerned with the polynomial continuous-time systems. To the best of the authors' knowledge, when it comes to polynomial discrete-time systems, no general result has been presented yet. However, closely related results can be found in [25,26]. In this work, we approximate nonlinear discrete-time systems by the Takagi–Sugeno (TS) fuzzy models, which are locally linear models connected by IF-THEN rules. To be specific, in [25,26], a convex solution to the H_∞ control problem is achieved by selecting a Lyapunov function of a rational form and introducing a transformation variable that is coupled with the system output matrices. In doing so, a change of variable is able to be applied to the bilinear term. In [25], a quadratic Lyapunov function is used to analyze the stability of such TS fuzzy systems. It is well known that quadratic Lyapunov functions are always inadequate to solve the problem of nonlinear systems or linear systems with structured uncertainties [27,28]. This is the main drawback of the quadratic Lyapunov function-based approach. Motivated by this fact, [26] uses a parameter-dependent Lyapunov function, and therefore this allows them to extend their approach to robust control synthesis for nonlinear systems. But by employing the parameter-dependent approach the consequence is that the solution to the H_∞ control problem is no longer convex. Hence, in this paper, to render a convex solution to the problem, the Lyapunov matrix is enforced to be diagonal in structure. This selection may lead to conservative results. Furthermore, in the TS fuzzy model, the premise variables are assumed to be bounded. In general, the premise variables are related to the state variables, which implies that the state variables have also to be bounded. This is one of the major drawbacks of the TS fuzzy model approach.

In this chapter, motivated by the aforementioned problems and the results in [25,26], the problem of H_∞ control using a static output feedback controller for a class of polynomial discrete-time systems is studied. In particular, we address the problem of H_∞ control in which both stability and a prescribed H_∞ performance are required to be fulfilled. To be specific, the polynomial discrete-time system is represented in the form of state-dependent linear-like form, and a state-dependent polynomial Lyapunov function is used to represent a Lyapunov candidate. Attention is focused on the design of the nonlinear static output feedback controller with an integrator to stabilize such discrete-time systems and to ensure that the prescribed level of the H_∞ performance is fulfilled. By incorporating the integrator into the controller structure, the original system can be transformed into the augmented system, and the Lyapunov matrix can be chosen to be dependent upon the original states only. Through this, the static output synthesis problem of polynomial discrete-time systems can be convexified in a less conservative way and can be solved computationally via SDP. The existence of the controller is given in terms of the solvability conditions of polynomial matrix inequalities (PMIs) formulated as SOS constraints and solved using SOSTOOLS [29]. It is important to note that the resulting controller gains are in the form of rational matrix functions of the system output matrices and the additional augmented state. The results are then directly extended to the robust H_∞ output feedback control with uncertainties.

In comparison with the existing method of a static output feedback controller design for nonlinear discrete-time systems, there are two features of our proposed approach that deserve attention:

1. By introducing an integrator into the controller structure, a less conservative result can be obtained. This is because the nonconvex term that exists between the Lyapunov function and the controller matrix due to the utilization of the state-dependent Lyapunov function can be convexified in a less conservative way than the available approaches. To be precise, by incorporating the integrator into the controller structure, the Lyapunov matrix need not be of a certain structure to render a convex solution. In contrast, to achieve this, [26] has to enforce the Lyapunov matrix to be in a diagonal form. This condition may give a conservative result. On the other hand, a predefined upper bound has been proposed in [30,22] to limit the effect of the nonconvex term. However, this predefined upper bound is hard to determine beforehand, and the results just only hold within a bounded region.
2. The Lyapunov function need not be of a rational form, and no additional transformation matrix is needed to apply a change-of-variable technique to the bilinear term as required in [26].

6.2 SYSTEM DESCRIPTION AND PROBLEM FORMULATION

We consider the following polynomial discrete-time system:

$$\begin{cases} x(k+1) = A(x(k))x(k) + B_u(x(k))u(k) + B_\omega(x(k))\omega(k), \\ z(k) = C_z(x(k))x(k) + D_{zu}(x(k))u(k), \\ y(k) = C_y(x(k))x(k), \end{cases} \quad (6.1)$$

where $x(k) \in R^n$ is the state vectors, $u(k) \in R^m$ is the input, $y(k)$ is the measured output, and $A(x(k))$, $B_u(x(k))$, $C_z(x(k))$, $D_{zu}(x(k))$, and $C_y(x(k))$ are polynomial matrices of appropriate dimensions. Meanwhile, $z(k)$ is a vector of output signals related to the performance of the control system, and $\omega(k)$ is the disturbance belonging to $L_2[0, \infty]$.

For the polynomial discrete-time system described in (6.1), we propose the output feedback controller

$$u(k) = K(y)y(k). \quad (6.2)$$

Before presenting the main result, we need the following lemmas.

Lemma 2. *[30,28] System* (6.1) *without disturbance, i.e.,* $\omega(k) = 0$, *is asymptotically stable if*

1. *there exist a positive definite symmetric polynomial matrix* $P(x(k))$ *and a polynomial matrix* $K(x(k))$ *such that*

$$\begin{bmatrix} P(x(k)) & * \\ P(x_+)A(x(k)) + P(x_+)B(x(k))K(x(k)) & P(x_+) \end{bmatrix} > 0, \quad (6.3)$$

   or

2. *there exist a positive definite symmetric polynomial matrix* $P(x(k))$, *a polynomial matrix* $K(x(k))$, *and polynomial slack matrix* $G(x(k))$ *such that*

$$\begin{bmatrix} P(x(k)) & * \\ G(x(k))A(x(k)) + G(x(k))B(x(k))K(x(k)) & G(x(k)) + G^T(x(k)) - P(x_+) \end{bmatrix} > 0. \quad (6.4)$$

Proof. A detailed proof can be found in [30,28] and hence is omitted. □

Lemma 3. *[31] Let* $F(x)$ *be an* $N \times N$ *symmetric polynomial matrix of degree* $2d$ *in* $x \in R^n$. *Furthermore, let* $Z(x)$ *be a column vector whose entries are all monomials in* x *of degree no greater than* d, *and let the following conditions be satisfied:*

1. *$F(x) \geq 0$ for all $x \in R^n$;*
2. *$v^T F(x)v$ is an SOS for $v \in R^N$;*
3. *There exists a positive semidefinite matrix Q such that $v^T F(x)v = (v \otimes Z(x))^T Q (v \otimes Z(x))$, with $\otimes$ denoting the Kronecker product.*

It is clear that if $F(x)$ is an SOS, then $F(x) \geq 0$, but the converse is generally not true. Furthermore, Statement (2) and Statement (3) are equivalent.

Remark 26. Lemma 2 shows that by utilizing a slack variable technique, i.e., introducing a slack polynomial matrix $G(x(k))$, a less conservative result can be obtained [30,28]. This is because a Lyapunov function can be decoupled from the system matrices, and therefore the controller design is independent of the Lyapunov matrix. The controller design is now dependent upon the slack polynomial matrix. By employing such a slack variable technique, the parameter-dependent Lyapunov function has been used for linear uncertain systems [28], and the state-dependent Lyapunov function has been utilized for polynomial systems [30]. Although the above lemma provides a solution for the state feedback control, the method has also been applied in the framework of static output feedback control designs as shown in [25,26]. Based on the slack variable technique and the state-dependent Lyapunov function, we attempt to derive a new method that provides a less conservative design procedure for designing a nonlinear H_∞ output feedback controller for polynomial discrete-time systems. This is delivered by incorporating an integrator into the controller structure.

6.3 MAIN RESULTS

We begin this section by highlighting the problem of designing an H_∞ output feedback controller for polynomial discrete-time systems when a state-dependent Lyapunov function is under consideration. Then, we propose a novel method to overcome that problem. Based on this novel method, a solution to the nonlinear H_∞ output feedback control problem is given. The results are subsequently extended to the robust H_∞ control with polytopic uncertainty.

6.3.1 Nonlinear H_∞ output feedback control

The following theorem provides sufficient conditions for the existence of a nonlinear static output feedback controller (6.2) for system (6.1) without disturbance, i.e., $\omega(k) = 0$.

Theorem 15. *System* (6.1) *without disturbance (i.e., $\omega(k) = 0$) is stabilizable asymptotically via the static output feedback control of the form* (6.2) *if there exist a symmetric polynomial matrix $P(x(k))$ and polynomial matrices $K(y)$ and $G(x(k))$ such that the following conditions are satisfied for* *all* $x \neq 0$:

$$P(x(k)) > 0, \tag{6.5}$$

$$\begin{bmatrix} P(x(k)) & * \\ G(x(k))A(x(k)) + G(x(k))B_u(x(k))K(y)C_y(x(k)) & G(x(k)) + G^T(x(k)) - P(x_+) \end{bmatrix} > 0. \tag{6.6}$$

Proof. Select a Lyapunov function of the form

$$V(x(k)) = x^T(k)P(x(k))x(k). \tag{6.7}$$

The difference of the Lyapunov function (6.7) along system (6.1) with (6.2) for $\omega(k) = 0$ is given by

$$\begin{aligned} \Delta V(x(k)) &= V(x(k+1)) - V(x(k)) < 0 \\ &= \big(A(x(k))x(k) + B_u(x(k))K(y)C_y(x(k))x(k)\big)^T P(x_+) \\ &\quad \big(A(x(k))x(k) + B_u(x(k))K(y)C_y(x(k))x(k)\big) \\ &\quad - x^T(k)P(x(k))x(k) \\ &= x^T(k)\big[\big(A(x(k)) + B_u(x(k))K(y)C_y(x(k))\big)^T P(x_+)\big(A(x(k)) \\ &\quad + B_u(x(k))K(y)C_y(x(k))\big) - P(x(k))\big]x(k). \end{aligned} \tag{6.8}$$

Suppose (6.6) holds. Then by Lemma 2 we have

$$\begin{bmatrix} P(x(k)) & * \\ P(x_+)A(x(k)) + P(x_+)B_u(x(k))K(y)C_y(x(k)) & P(x_+) \end{bmatrix} > 0. \tag{6.9}$$

Next, multiplying (6.9) on the left by $diag[I, P^{-1}(x_+)]$ and on the right by $diag[I, P^{-1}(x_+)]^T$, we get

$$\begin{bmatrix} P(x(k)) & * \\ A(x(k)) + B_u(x(k))K(y)C_y(x(k)) & P^{-1}(x_+) \end{bmatrix} > 0. \tag{6.10}$$

Then, by applying the Schur complement to (6.10) we obtain

$$\begin{aligned} &\big[(A(x(k)) + B_u(x(k))K(y)C_y(x(k)))^T P(x_+)\big(A(x(k)) \\ &\quad + B_u(x(k))K(y)C_y(x(k))\big) - P(x(k))\big] < 0. \end{aligned} \tag{6.11}$$

Knowing that (6.11) holds, we have $\Delta V(x(k)) < 0$ for $x \neq 0$, which implies that system (6.1) with (6.2) is asymptotically stable. This completes the proof. □

It is worth mentioning that the conditions given in Theorem 15 are in terms of state-dependent PMIs. Thus, solving this inequality is computationally hard because we need to solve an infinite set of state-dependent PMIs. To relax these conditions, we utilize the SOS decomposition approach and semidefinite programming as described in [32,33], where the conditions in Theorem 15 can be solved by parameterizing $P(x(k))$ and $K(y)$ in a proper polynomial form. Moreover, to render conditions given in Theorem 15 into tractable SOS conditions, it is often necessary to include some SOS constraints, i.e., $\epsilon > 0$ [33]. Therefore the static output feedback stabilization conditions given in Theorem 15 can be modified into SOS conditions given by the following proposition.

Proposition 3. *System* (6.1) *without disturbance (i.e.,* $\omega(k) = 0$*) is asymptotically stable via static output feedback controller* (6.2) *if there exist a symmetric polynomial matrix* $P(x(k))$*, polynomial matrices* $K(y)$ *and* $G(x(k))$*, and positive constants* $\epsilon_1 > 0$ *and* $\epsilon_2 > 0$ *such that the following conditions hold for all* $x \neq 0$*:*

$$v_1^T [P(x(k)) - \epsilon_1 I] v_1 \quad \text{is an SOS,} \tag{6.12}$$

$$v_2^T [M(x(k)) - \epsilon_2 I] v_2 \quad \text{is an SOS,} \tag{6.13}$$

where

$$M(x) = \begin{bmatrix} P(x) & * \\ G(x)A(x) + G(x)B_u(x)K(y)C_y(x) & G(x) + G^T(x) - P(x_+) \end{bmatrix}, \tag{6.14}$$

and v_1 *and* v_2 *are free vectors of appropriate dimensions.*

Proof. The proof follows directly from that of Theorem 15. In addition, knowing that inequalities (6.5) and (6.6) are in symmetric form. Therefore, if the Proposition 3 holds, this implies that Theorem 15 is true. The proof ends. □

Remark 27. Unfortunately, Proposition 3 cannot be solved easily by SDP because:

1. A change-of-variable technique cannot be applied directly to the bilinear term $G(x(k))B_u(x(k))K(y)$ due to the existence of $B_u(x(k))$ in between of the additional slack variable matrix $G(x(k))$ and the controller matrix $K(y)$. One possible way to solve this problem is by imposing the

Lyapunov function to be of rational form and introducing a transformation variable, T such that $C_y T = [I, 0]$ [25,26]. Then, by enforcing the slack variable matrix to be of a certain form, the change-of-variable technique can be applied to the bilinear term accordingly. However, the combination of C_y and T is not unique, and hence it is difficult to choose a suitable candidate for T. In light of this method, it is not hard to see that the change-of-variable technique can be applied to the $G(x(k))B_u(x(k))K(y)$ of (6.14) easily by forcing the $B_u(x(k))$ to be $[0, 1]^T$ and choosing the $G(x(k))$ to be of a certain form (as shown in [26]). In doing so, a change-of-variable can be applied to the bilinear term, but the result becomes more conservative because the input matrix $B_u(x(k))$ must always be $[0, 1]^T$.

2. The terms in $P(x_+)$ are not jointly convex. This is true because if we expand it, then we have

$$\begin{aligned} P(x_+) &= P(x(k+1)) \\ &= P(A(x(k))x(k) + B_u(x(k))K(y)C(x(k))x(k)). \end{aligned} \quad (6.15)$$

It is obvious the terms in (6.15) are not jointly convex; hence, it is hard to search for these values simultaneously, and therefore it is hard to solve (6.15), which is equivalent to solving some bilinear matrix inequalities (BMIs). It has been shown in [30] that one possible way to convexify this problem is by introducing a predefined upper bound to limit the effect of the nonconvex term. This predefined upper bound, however, is hard to be determined beforehand, and the closed-loop stability can only be guaranteed within a bound region. Another possible solution is by selecting a Lyapunov function to be of a quadratic form as applied in [25]. But it is well known that a quadratic Lyapunov function is always inadequate to solving nonlinear systems.

Motivated by the above-mentioned problems and the results in [25,26], we introduce an integrator into the static output controller structure. In doing this, the problems mentioned in Remark 27 can be resolved in a less conservative way than the available approaches.

Our proposed controller with an integrator is given as follows:

$$\begin{aligned} x_c(k+1) &= x_c(k) + A_c(y, x_c), \\ u(k) &= x_c(k), \end{aligned} \quad (6.16)$$

where x_c is the controller state, $u(k)$ is the input to the system, and $A_c(y, x_c)$ is the input function of the integrator.

System (6.1) with controller (6.16) can be described as follows:

$$\begin{cases} \hat{x}(k+1) = \hat{A}(\hat{x}(k))\hat{x}(k) + \hat{B}_u(\hat{x}(k))A_c(y, x_c) + \hat{B}_\omega(\hat{x}(k))\omega(k), \\ \quad z(k) = \hat{C}_z(\hat{x}(k))\hat{x}(k), \\ \quad y(k) = \hat{C}_y(\hat{x}(k))\hat{x}(k), \end{cases} \tag{6.17}$$

where

$$\begin{aligned} \hat{A}(\hat{x}(k)) &= \begin{bmatrix} A(x(k)) & B(x(k)) \\ 0 & 1 \end{bmatrix}, \quad \hat{B}_u(\hat{x}(k)) = \begin{bmatrix} 0 \\ 1 \end{bmatrix}, \\ \hat{B}_\omega(\hat{x}(k)) &= \begin{bmatrix} B_\omega(x(k)) \\ 0 \end{bmatrix}, \quad \hat{C}_z(\hat{x}(k)) = \begin{bmatrix} C_z(x(k)) & D_{zu}(x(k)) \end{bmatrix}, \\ \hat{C}_y(\hat{x}(k)) &= \begin{bmatrix} C_y(x(k)) & 0 \end{bmatrix}, \quad \hat{x} = \begin{bmatrix} x(k) \\ x_c(k) \end{bmatrix}. \end{aligned} \tag{6.18}$$

Next, we assume $A_c(y, x_c)$ to be of the form

$$A_c(k) = \hat{K}(y, x_c)y, \tag{6.19}$$

where $\hat{K}(y, x_c)$ is a polynomial matrix of dimensions $(n+1) \times 1$, n is the number of original system states. Therefore, (6.17) can be rewritten as follows:

$$\begin{Bmatrix} \hat{x}(k+1) = \hat{A}(\hat{x}(k))\hat{x}(k) + \hat{B}_u(\hat{x}(k))\hat{K}(y, x_c)y + \hat{B}_\omega(\hat{x}(k))\omega(k), \\ \quad z(k) = \hat{C}_z(\hat{x}(k))\hat{x}(k), \\ \quad y(k) = \hat{C}_y(\hat{x}(k))\hat{x}(k). \end{Bmatrix} \tag{6.20}$$

Here, the objective is to design a nonlinear output feedback controller of the form (6.16) such that for a given prescribed H_∞ performance $\gamma > 0$,

$$\|z(k)\|_{[0,\infty]} \leq \gamma_2 \|\omega(k)\|_{[0,\infty]}, \tag{6.21}$$

and system (6.1) with (6.16) is globally asymptotically stable.

Now, we are ready to present our main result. Sufficient conditions for the existence of our proposed controller (6.16) for system (6.1) without disturbance, i.e., $\omega(k) = 0$, are given in the following corollary.

Corollary 8. *System* (6.1) *without disturbance (i.e.,* $\omega(k) = 0$*) is asymptotically stable via the nonlinear output feedback controller* (6.16) *if there exist a symmetric polynomial matrix* $\hat{P}(x(k))$*, a polynomial function* $L_{31}(y, x_c)$*,* *a polynomial matrix $\hat{G}(\hat{x}(k))$, and positive constants $\epsilon_1 > 0$ and $\epsilon_2 > 0$ such that the following conditions hold for all $x \neq 0$:*

$$v_3^T[\hat{P}(x(k)) - \epsilon_1 I]v_3 \quad \text{is an SOS,} \tag{6.22}$$

$$v_4^T\left[M(\hat{x}(k)) - \epsilon_2 I\right]v_4 \quad \text{is an SOS,} \tag{6.23}$$

where v_3 and v_4 are free vectors of appropriate dimensions, and

$$M(\hat{x}(k)) = \begin{bmatrix} \hat{P}(x(k)) & * \\ \hat{G}(\hat{x}(k))\hat{A}(\hat{x}(k)) + \hat{L}(y, x_c)\hat{C}_y(\hat{x}(k)) & \hat{G}(\hat{x}(k)) + \hat{G}^T(\hat{x}(k)) - \hat{P}(x_+) \end{bmatrix} \tag{6.24}$$

with

$$\hat{G}(\hat{x}(k)) = \begin{bmatrix} G_{11}(\hat{x}(k)) & G_{12}(\hat{x}(k)) & 0 \\ G_{21}(\hat{x}(k)) & G_{22}(\hat{x}(k)) & 0 \\ G_{31}(\hat{x}(k)) & G_{32}(\hat{x}(k)) & G_{33}(y, x_c) \end{bmatrix},$$

$$\hat{L}(y, x_c) = \begin{bmatrix} 0 \\ 0 \\ L_{31}(y, x_c) \end{bmatrix}. \tag{6.25}$$

Moreover, the nonlinear output feedback controller is given by

$$\begin{aligned} x_c(k+1) &= x_c(k) + A_c(y, x_c), \\ u(k) &= x_c(k), \end{aligned} \tag{6.26}$$

where

$$\begin{aligned} A_c(y, x_c) &= \hat{K}(y, x_c)\hat{C}_y(\hat{x}(k))\hat{x}(k) \\ \text{with} \quad \hat{K}(y, x_c) &= L_{31}(y, x_c)G_{33}^{-1}(y, x_c). \end{aligned} \tag{6.27}$$

Proof. We select a Lyapunov function of the form

$$\hat{V}(\hat{x}(k)) = \hat{x}^T(k)\hat{P}(x(k))\hat{x}(k), \tag{6.28}$$

and we let

$$\hat{L}(\hat{x}(k)) = \begin{bmatrix} 0 \\ 0 \\ L_{31}(y, x_c) \end{bmatrix} = \begin{bmatrix} 0 \\ 0 \\ G_{33}(y, x_c)\hat{K}(y, x_c) \end{bmatrix}$$

$$
\begin{aligned}
&= \begin{bmatrix} G_{11}(\hat{x}(k)) & G_{12}(\hat{x}(k)) & 0 \\ G_{21}(\hat{x}(k)) & G_{22}(\hat{x}(k)) & 0 \\ G_{31}(\hat{x}(k)) & G_{32}(\hat{x}(k)) & G_{33}(y, x_c) \end{bmatrix} \begin{bmatrix} 0 \\ 0 \\ \hat{K}(y, x_c) \end{bmatrix} \\
&= \begin{bmatrix} G_{11}(\hat{x}(k)) & G_{12}(\hat{x}(k)) & 0 \\ G_{21}(\hat{x}(k)) & G_{22}(\hat{x}(k)) & 0 \\ G_{31}(\hat{x}(k)) & G_{32}(\hat{x}(k)) & G_{33}(y, x_c) \end{bmatrix} \begin{bmatrix} 0 \\ 0 \\ 1 \end{bmatrix} \hat{K}(y, x_c) \\
&= \hat{G}(\hat{x}(k))\hat{B}_u(\hat{x}(k))\hat{K}(y, x_c). \qquad (6.29)
\end{aligned}
$$

Then, by applying a similar technique of the proof shown in Theorem 15, it is trivial to show that we have the following inequalities:

$$
\begin{aligned}
&\hat{P}(x(k)) > 0 \quad \text{and} \\
&M(\hat{x}(k)) > 0, \qquad (6.30)
\end{aligned}
$$

where $M(\hat{x}(k))$ is as described in (6.24). Furthermore, by utilizing Proposition 3 we can show that if inequalities (6.22)–(6.23) are SOS, then inequalities (6.30) hold. Hence, the proof is completed. □

Remark 28. Note that our proposed controller $\hat{K}(y, x_c)$ depends upon the augmented system output matrices. This is to ensure that a feasible solution to the problem can be obtained.

Remark 29. **1.** To allow us to apply a change-of-variable technique to the bilinear term $\hat{G}(\hat{x}(k))\hat{B}_u(\hat{x}(k))\hat{K}(y, x_c)$, we follow the method shown in [25,26], but our method provides a less conservative way because the Lyapunov function need not be of a rational form and no transformation matrix is required. This is due to the fact that our $\hat{B}_u(\hat{x}(k))$ is always in the vector of $[0, 0, 1]^T$. In addition, the control matrix $B_u(x(k))$ is now governed in the system matrices, and hence the Lyapunov matrix does need not be of a special form to render a convex solution, that is, the Lyapunov function need not depend upon the system states whose corresponding rows in the control matrix are zeros. Therefore, our method produces a more general result than [30].

2. The term $P(x_+)$ is now in a convex form. To explain this, refer to our proposed Lyapunov matrix (6.28), where $\hat{P}(x(k))$ only depends upon the original system matrices. Hence, expanding the term $P(x_+)$, we have

$$
\begin{aligned}
\hat{P}(x_+) &= \hat{P}(x(k+1)) \\
&= \hat{P}(A(x(k))x(k) + B(x(k))u(k)) \\
&= \hat{P}(A(x(k))x(k) + B(x(k))x_c(k)). \qquad (6.31)
\end{aligned}
$$

Now it is not hard to see the terms in (6.31) are jointly convex because $x_c(k)$ is an augmented state, which allows us to possibly solve Corollary 8 via SDP.

Theorem 16. *Given a prescribed H_∞ performance $\gamma > 0$, system (6.1) is asymptotically stable via the nonlinear output feedback controller (6.16) with H_∞ performance (6.21) if there exist a symmetric polynomial matrix $\hat{P}(x(k))$, a polynomial function $L_{31}(y, x_c)$, and a polynomial matrix $\hat{G}(\hat{x}(k))$ such that the following conditions hold for all $x \neq 0$:*

$$\hat{P}(x(k)) > 0, \tag{6.32}$$

$$M_2(\hat{x}(k)) > 0, \tag{6.33}$$

where

$$M_2(\hat{x}(k)) = \begin{bmatrix} \hat{P}(x(k)) & * & * & * \\ 0 & \gamma^2 I & * & * \\ \hat{G}(\hat{x}(k))\hat{A}(\hat{x}(k)) + \hat{L}(y, x_c)\hat{C}_y(\hat{x}(k)) & \hat{B}_\omega(\hat{x}(k)) & \hat{G}(\hat{x}(k)) + \hat{G}^T(\hat{x}(k)) - \hat{P}(x_+) & * \\ \hat{C}_z(\hat{x}(k)) & 0 & 0 & I \end{bmatrix} \tag{6.34}$$

with

$$\hat{G}(\hat{x}(k)) = \begin{bmatrix} G_{11}(\hat{x}(k)) & G_{12}(\hat{x}(k)) & 0 \\ G_{21}(\hat{x}(k)) & G_{22}(\hat{x}(k)) & 0 \\ G_{31}(\hat{x}(k)) & G_{32}(\hat{x}(k)) & G_{33}(y, x_c) \end{bmatrix},$$

$$\hat{L}(y, x_c) = \begin{bmatrix} 0 \\ 0 \\ L_{31}(y, x_c) \end{bmatrix}. \tag{6.35}$$

Moreover, the nonlinear output feedback controller is given by

$$\begin{aligned} x_c(k+1) &= x_c(k) + A_c(y, x_c), \\ u(k) &= x_c(k), \end{aligned} \tag{6.36}$$

where

$$\begin{aligned} A_c(y, x_c) &= \hat{K}(y, x_c)\hat{C}_y(\hat{x}(k))\hat{x}(k) \\ \textit{with} \quad \hat{K}(y, x_c) &= L_{31}(y, x_c)G_{33}^{-1}(y, x_c). \end{aligned} \tag{6.37}$$

Proof. Based on the Lyapunov function given in (6.28), the $V(x(k+1)) - V(x(k))$ along (6.20) is given by

$$\Delta\hat{V}(\hat{x}(k)) = \hat{V}(\hat{x}(k+1)) - \hat{V}(\hat{x}(k))$$

$$= \hat{x}^T(k+1)\hat{P}(x_+)\hat{x}(k+1) - \hat{x}^T(k)\hat{P}(x(k))\hat{x}(k)$$
$$= \big(\hat{A}(\hat{x}(k))\hat{x}(k) + \hat{B}_u(\hat{x}(k))\hat{K}(y, x_c)\hat{C}_y(\hat{x}(k))\hat{x}(k)$$
$$+ B_\omega(x(k))\omega(k)\big)^T \hat{P}(x_+)\big(\hat{A}(\hat{x}(k))\hat{x}(k) + \hat{B}_u(\hat{x}(k))\hat{K}(y, x_c)\hat{C}_y(\hat{x}(k))\hat{x}(k)$$
$$+ B_\omega(x(k))\omega(k)\big) - \hat{x}^T(k)\hat{P}(x(k))\hat{x}(k). \tag{6.38}$$

Furthermore, adding and subtracting $-z^T(k)z(k) + \gamma^2\omega^T(k)\omega(k)$ to and from (6.38) result in

$$\Delta\hat{V}(\hat{x}(k)) = \big(\hat{A}(\hat{x}(k))\hat{x}(k) + \hat{B}_u(\hat{x}(k))\hat{K}(y, x_c)\hat{C}_y(\hat{x}(k))\hat{x}(k)$$
$$+ B_\omega(x(k))\omega(k)\big)^T \hat{P}(x_+)\big(\hat{A}(\hat{x}(k))\hat{x}(k) + \hat{B}_u(\hat{x}(k))\hat{K}(y, x_c)\hat{C}_y(\hat{x}(k))\hat{x}(k)$$
$$+ B_\omega(x(k))\omega(k)\big) - \hat{x}^T(k)\hat{P}(x(k))\hat{x}(k) - z^T(k)z(k) + \gamma^2\omega^T(k)\omega(k)$$
$$+ z^T(k)z(k) - \gamma^2\omega^T(k)\omega(k). \tag{6.39}$$

Knowing that $z(k) = \hat{C}_z(\hat{x}(k)\hat{x}(k))$, (6.39) becomes

$$\Delta\hat{V}(\hat{x}(k))$$
$$= \big(\hat{A}(\hat{x}(k))\hat{x}(k) + \hat{B}_u(\hat{x}(k))\hat{K}(y, x_c)\hat{C}_y(\hat{x}(k))\hat{x}(k)$$
$$+ B_\omega(x(k))\omega(k)\big)^T \hat{P}(x_+)\big(\hat{A}(\hat{x}(k))\hat{x}(k) + \hat{B}_u(\hat{x}(k))\hat{K}(y, x_c)\hat{C}_y(\hat{x}(k))\hat{x}(k)$$
$$+ B_\omega(x(k))\omega(k)\big) - \hat{x}^T(k)\hat{P}(x(k))\hat{x}(k) + (\hat{C}_z(\hat{x}(k))\hat{x}(k))^T(\hat{C}_z(\hat{x}(k))\hat{x}(k))$$
$$- \gamma^2\omega^T(k)\omega(k) - z^T(k)z(k) + \gamma^2\omega^T(k)\omega(k). \tag{6.40}$$

Now, (6.40) can be rewritten as follows:

$$\Delta\hat{V}(\hat{x}(k)) = \hat{X}^T(k)\Omega(\hat{x}(k))\hat{X}(k) - z^T(k)z(k) + \gamma^2\omega^T(k)\omega(k), \tag{6.41}$$

where

$$\Omega(\hat{x}(k)) = \phi_1(\hat{x}(k))^T P(x_+)\phi_1(\hat{x}(k)) + \phi_2(\hat{x}(k))^T\phi_2(\hat{x}(k)) - \Xi$$

with

$$\phi_1(\hat{x}(k)) = \begin{bmatrix} \hat{A}(\hat{x}(k)) + \hat{B}_u(\hat{x}(k))\hat{K}(y, x_c)\hat{C}_y(\hat{x}(k)) & \hat{B}_\omega(\hat{x}(k)) \end{bmatrix},$$
$$\phi_2(\hat{x}(k)) = \begin{bmatrix} \hat{C}_z(\hat{x}(k)) & 0 \end{bmatrix}, \quad \hat{X}(k) = \begin{bmatrix} x(k) \\ \omega(k) \end{bmatrix}, \quad \Xi = \begin{bmatrix} \hat{P}(x(k)) & 0 \\ 0 & \gamma^2 \end{bmatrix}.$$

Now, we need to show that $\hat{X}^T(k)\Omega(\hat{x}(k))\hat{X}(k) < 0$. Suppose that (6.33) is feasible. Then from the block $(3, 3)$ of (6.34) we have $\hat{G}(\hat{x}(k)) + \hat{G}^T(\hat{x}(k)) > \hat{P}(x_+) > 0$. This implies that $\hat{G}(\hat{x}(k))$ is nonsingular, and since $\hat{P}(x_+)$ is positive definite, we have

$$\left(\hat{P}(x_+) - \hat{G}(\hat{x}(k))\right)\hat{P}^{-1}(x_+)\left(\hat{P}(x_+) - \hat{G}(\hat{x}(k))\right)^T > 0, \tag{6.42}$$

and therefore

$$\hat{G}(\hat{x}(k))\hat{P}^{-1}(x_+)\hat{G}^T(\hat{x}(k)) \geq \hat{G}(\hat{x}(k)) + \hat{G}^T(\hat{x}(k)) - \hat{P}(x_+). \tag{6.43}$$

This immediately gives

$$\begin{bmatrix} \hat{P}(x(k)) & * & * & * \\ 0 & \gamma^2 I & * & * \\ \hat{G}(\hat{x}(k))\hat{A}(\hat{x}(k)) + \hat{L}(\hat{x}(k))\hat{C}_y(x(k)) & \hat{B}_\omega(\hat{x}(k)) & \hat{G}(\hat{x}(k))\hat{P}^{-1}(x_+)\hat{G}^T(\hat{x}(k)) & * \\ \hat{C}_z(\hat{x}(k)) & 0 & 0 & I \end{bmatrix} > 0. \tag{6.44}$$

On the other hand, from (6.35) and (6.37) and from the fact that $\hat{B}_u(\hat{x}(k))$ is always $[0, 0, 1]^T$ we have

$$\begin{aligned} \hat{L}(\hat{x}(k)) &= \begin{bmatrix} 0 \\ 0 \\ L_{31}(y, x_c) \end{bmatrix} = \begin{bmatrix} 0 \\ 0 \\ S_{33}(y, x_c)\hat{K}(y, x_c) \end{bmatrix} \\ &= \begin{bmatrix} G_{11}(\hat{x}(k)) & G_{12}(\hat{x}(k)) & 0 \\ G_{21}(\hat{x}(k)) & g_{22}(\hat{x}(k)) & 0 \\ G_{31}(\hat{x}(k)) & G_{32}(\hat{x}(k)) & G_{33}(y, x_c) \end{bmatrix} \begin{bmatrix} 0 \\ 0 \\ \hat{K}(y, x_c) \end{bmatrix} \\ &= \begin{bmatrix} G_{11}(\hat{x}(k)) & G_{12}(\hat{x}(k)) & 0 \\ G_{21}(\hat{x}(k)) & G_{22}(\hat{x}(k)) & 0 \\ G_{31}(\hat{x}(k)) & G_{32}(\hat{x}(k)) & G_{33}(y, x_c) \end{bmatrix} \begin{bmatrix} 0 \\ 0 \\ 1 \end{bmatrix} \hat{K}(y, x_c) \\ &= \hat{G}(\hat{x}(k))\hat{B}_u(\hat{x}(k))\hat{K}_y(y, x_c). \end{aligned} \tag{6.45}$$

Next, multiplying (6.44) on the right by $diag[I, I, G^{-1}(\hat{x}(k)), I]^T$ and on the left by $diag[I, I, G^{-1}(\hat{x}(k)), I]$, we get

$$\begin{bmatrix} \hat{P}(x(k)) & * & * & * \\ 0 & \gamma^2 I & * & * \\ \hat{A}(\hat{x}(k)) + \hat{B}_u(x(k))\hat{K}(y, x_c)\hat{C}_y(x(k)) & \hat{B}_\omega(\hat{x}(k)) & \hat{P}^{-1}(x_+) & * \\ \hat{C}_z(\hat{x}(k)) & 0 & 0 & I \end{bmatrix} > 0 \tag{6.46}$$

and, similarly,

$$\begin{bmatrix} \Xi & \phi_1(\hat{x}(k))^T & \phi_2(\hat{x}(k))^T \\ \phi_1(\hat{x}(k)) & \hat{P}^{-1}(x_+) & 0 \\ \phi_2(\hat{x}(k)) & 0 & I \end{bmatrix} > 0. \tag{6.47}$$

Applying the Schur complement to (6.47), we have

$$\left(\phi_1(\hat{x}(k))^T P(x_+)\phi_1(\hat{x}(k)) + \phi_2(\hat{x}(k))^T \phi_2(\hat{x}(k)) - \Xi\right) < 0. \tag{6.48}$$

Now, knowing that (6.48) holds, from (6.41) we have

$$\Delta\hat{V}(\hat{x}(k)) < -z^T(k)z(k) + \gamma^2\omega^T(k)\omega(k). \tag{6.49}$$

Furthermore, the summation from 0 to ∞ yields

$$\hat{V}(x(\infty)) - \hat{V}(\hat{x}(0)) \le -\sum_{k=0}^{\infty} z^T(k)z(k) + \sum_{k=0}^{\infty} \gamma^2\omega^T(k)\omega(k), \tag{6.50}$$

and since $\hat{V}(\hat{x}(0)) = 0$ and $\hat{V}(\hat{x}(\infty)) \ge 0$, we obtain

$$\sum_{k=0}^{\infty} z^T(k)z(k) \le \gamma^2 \sum_{k=0}^{\infty} \omega^T(k)\omega(k). \tag{6.51}$$

Thus, (6.21) holds, and the H_∞ performance is fulfilled.

To prove that system (6.1) with (6.16) is asymptotically stable, we set the disturbance $\omega(k) = 0$. Asymptotic stability for such polynomial discrete-time systems is already shown in Corollary 8. This completes the proof. □

However, solving Theorem 16 is hard because we need to solve an infinite set of state-dependent PMIs. To relax these conditions, we utilize the SOS decomposition approach [32], and therefore the conditions given in Theorem 16 can be converted into SOS conditions given in the following corollary.

Corollary 9. *Given a prescribed H_∞ performance $\gamma > 0$, system* (6.1) *is asymptotically stable via the nonlinear output feedback controller* (6.16) *with H_∞ performance* (6.21) *if there exist a symmetric polynomial matrix $\hat{P}(x(k))$, a polynomial function $L_{31}(y, x_c)$, a polynomial matrix $\hat{G}(\hat{x}(k))$, and constants $\epsilon_1 > 0$ and $\epsilon_2 > 0$ such that the following conditions hold for all $x \neq 0$:*

$$v_5^T[\hat{P}(x(k)) - \epsilon_1 I]v_5 \quad \text{is an SOS}, \tag{6.52}$$

$$v_6^T\left[M_2(\hat{x}(k)) - \epsilon_2 I\right]v_6 \quad \text{is an SOS}, \tag{6.53}$$

where, υ_5 and υ_6 are free vectors of appropriate dimensions, and

$$M_2(\hat{x}(k)) = \begin{bmatrix} \hat{P}(x) & * & * & * \\ 0 & \gamma^2 I & * & * \\ \hat{G}(\hat{x})\hat{A}(\hat{x}) + \hat{L}(y, x_c)\hat{C}_y(\hat{x}) & \hat{B}_\omega(\hat{x}) & \hat{G}(\hat{x}) + \hat{G}^T(\hat{x}) - \hat{P}(x_+) & * \\ \hat{C}_z(\hat{x}) & 0 & 0 & I \end{bmatrix} \quad (6.54)$$

with

$$\hat{G}(\hat{x}(k)) = \begin{bmatrix} G_{11}(\hat{x}(k)) & G_{12}(\hat{x}(k)) & 0 \\ G_{21}(\hat{x}(k)) & G_{22}(\hat{x}(k)) & 0 \\ G_{31}(\hat{x}(k)) & G_{32}(\hat{x}(k)) & G_{33}(y, x_c) \end{bmatrix},$$
$$\hat{L}(y, x_c) = \begin{bmatrix} 0 \\ 0 \\ L_{31}(y, x_c) \end{bmatrix}. \quad (6.55)$$

Moreover, the nonlinear output feedback controller is given by

$$\begin{aligned} x_c(k+1) &= x_c(k) + A_c(y, x_c), \\ u(k) &= x_c(k), \end{aligned} \quad (6.56)$$

where

$$\begin{aligned} A_c(y, x_c) &= \hat{K}(y, x_c)\hat{C}_y(\hat{x}(k))\hat{x}(k) \\ \text{with} \quad \hat{K}(y, x_c) &= L_{31}(y, x_c)G_{33}^{-1}(y, x_c). \end{aligned} \quad (6.57)$$

Proof. Proof for this theorem follows directly by combining the proofs of Theorem 16 and Lemma 3. □

The advantages of formulating the conditions of the nonlinear output feedback problem with prescribed H_∞ performance γ in the form of Corollary 9 are twofold:

1. A less conservative design procedure can be achieved (refer to Remark 29).
2. The output feedback controller is decoupled from the Lyapunov function, and hence the controller design can be performed in a more relaxed way because it is independent from the Lyapunov matrix.

6.3.2 Robust nonlinear H_∞ output feedback control

In this section, we consider nonlinear systems with parametric uncertainties or norm-bounded uncertainties.

6.3.2.1 Parametric uncertainties

Consider the system

$$\begin{aligned} x(k+1) =& A(x(k),\theta)x(k) + B_u(x(k),\theta)u(k) + B_\omega(x(k),\theta)\omega(k), \\ z(k) =& C_z(x(k),\theta)x(k) + D_{zu}(x(k),\theta)u(k), \\ y(k) =& C_y(x(k),\theta)x(k), \end{aligned} \tag{6.58}$$

where the matrices $\cdot(x(k),\theta)$ are defined as

$$\begin{aligned} A(x(k),\theta) &= \sum_{i=1}^{q} A_i(x(k))\theta_i, \quad & B_u(x(k),\theta) &= \sum_{i=1}^{q} B_i(x(k))\theta_i, \\ C_z(x(k),\theta) &= \sum_{i=1}^{q} C_{zi}(x(k))\theta_i, \quad & D_{zu}(x(k),\theta) &= \sum_{i=1}^{q} D_{zui}(x(k))\theta_i, \\ C_y(x(k)) &= \sum_{i=1}^{q} C_y(x(k))\theta_i, \quad & B_\omega(x(k),\theta) &= \sum_{i=1}^{q} B_{\omega i}(x(k))\theta_i. \end{aligned} \tag{6.59}$$

$\theta = \left[\theta_1, \ldots, \theta_q\right]^T \in \mathbb{R}^q$ is the vector of constant uncertainty satisfying

$$\theta \in \Theta \triangleq \left\{ \theta \in \mathbb{R}^q : \theta_i \geq 0, i = 1, \ldots, q, \sum_{i=1}^{q} \theta_i = 1 \right\}. \tag{6.60}$$

We further define the following parameter-dependent Lyapunov function

$$\hat{V}(\hat{x}(k)) = \hat{x}^T(k)\Big(\sum_{i=1}^{q} \hat{P}_i(x(k))\theta_i\Big)^{-1}\hat{x}(k). \tag{6.61}$$

With the results from the previous section, the main result for robust H_∞ control problem can be proposed directly and given by the following corollary.

Corollary 10. *Given a prescribed H_∞ performance $\gamma > 0$ and constants $\epsilon_1 > 0$ and $\epsilon_2 > 0$ for $x \neq 0$ and $i = 1, \ldots, q$, system* (6.58) *with the nonlinear feedback controller* (6.16) *is asymptotically stable with H_∞ performance* (6.21) *for $x \neq 0$ if there exist a common polynomial matrix*

$\hat{G}(\hat{x}(k))$, a common polynomial function $\hat{L}_{31}(\hat{x}, y)$, and symmetric polynomial matrices, $\hat{P}_i(x(k))$ such that the following conditions are satisfied for all $x \neq 0$:

$$v_7^T[\hat{P}_i(x(k)) - \epsilon_1 I]v_7 \quad \text{is an SOS,} \tag{6.62}$$

$$v_8^T\left[M_3(\hat{x}(k)) - \epsilon_2 I\right]v_8 \quad \text{is an SOS,} \tag{6.63}$$

where, v_7 and v_8 are free vectors of appropriate dimensions, and

$$M_3(\hat{x}(k)) = \begin{bmatrix} \hat{P}_i(x(k)) & * & * & * \\ 0 & \gamma^2 I & * & * \\ \hat{G}(\hat{x})\hat{A}_i(\hat{x}) + \hat{L}(y, x_c)\hat{C}_{yi}(\hat{x}) & \hat{B}_{\omega i}(\hat{x}) & \hat{G}(\hat{x}) + \hat{G}^T(\hat{x}) - \hat{P}_i(x_+) & * \\ \hat{C}_{zi}(\hat{x}) & 0 & 0 & I \end{bmatrix} \tag{6.64}$$

with

$$\hat{G}(\hat{x}(k)) = \begin{bmatrix} G_{11}(\hat{x}(k)) & G_{12}(\hat{x}(k)) & 0 \\ G_{21}(\hat{x}(k)) & G_{22}(\hat{x}(k)) & 0 \\ G_{31}(\hat{x}(k)) & G_{32}(\hat{x}(k)) & G_{33}(y, x_c) \end{bmatrix},$$
$$\hat{L}(y, x_c) = \begin{bmatrix} 0 \\ 0 \\ L_{31}(y, x_c) \end{bmatrix}. \tag{6.65}$$

Moreover, the nonlinear output feedback controller is given by

$$\begin{aligned} x_c(k+1) &= x_c(k) + A_c(y, x_c), \\ u(k) &= x_c(k), \end{aligned} \tag{6.66}$$

where

$$\begin{aligned} A_c(y, x_c) &= \hat{K}(y, x_c)\hat{C}_y(\hat{x}(k))\hat{x}(k) \\ \text{with} \quad \hat{K}(y, x_c) &= L_{31}(y, x_c)G_{33}^{-1}(y, x_c). \end{aligned} \tag{6.67}$$

6.3.2.2 Norm-bounded uncertainties

Consider the following uncertain polynomial discrete-time system:

$$\begin{cases} x(k+1) = A(x(k))x(k) + \Delta A(x(k))x(k) + B_u(x(k))u(k) \\ \qquad + \Delta B_u(x(k))u(k) + B_\omega(x(k))\omega(k), \\ z(k) = C_z(x(k))x(k) + D_{zu}(x(k))u(k), \\ y(k) = C_y(x(k))x(k), \end{cases} \tag{6.68}$$

where $x(k) \in R^n$ is the state vector, $u(k) \in R^m$ is the input, $y(k)$ is the measured output, $A(x(k))$, $B_u(x(k))$, $C_z(x(k))$, $D_{zu}(x(k))$, and $C_y(x(k))$ are polynomial matrices of appropriate dimensions, $z(k)$ is a vector of output signals related to the performance of the control system, and $\omega(k)$ is the disturbance belonging to $L_2[0, \infty]$. Meanwhile, $\Delta A(x(k))$ and $\Delta B_u(x(k))$ represent the uncertainties in the system and satisfy the following assumption.

Assumption. The parameter uncertainties considered here are norm-bounded and given as follows:

$$\begin{bmatrix} \Delta A(x(k)) & \Delta B_u(x(k)) \end{bmatrix} = H(x(k))F(x(k)) \begin{bmatrix} E_1(x(k)) & E_2(x(k)) \end{bmatrix}, \tag{6.69}$$

where $H(x(k))$, $E_1(x(k))$, and $E_2(x(k))$ are known polynomial matrices of appropriate dimensions, and $F(x(k))$ is an unknown matrix function satisfying

$$\left\| F^T(x(k))F(x(k)) \right\| \leq I. \tag{6.70}$$

□

To ensure a convex solution of $P(x(k+1))$, we propose the following nonlinear feedback controller:

$$\begin{cases} x_c(k+1) = x_c(k) + A_c(y, x_c), \\ u(k) = x_c(k), \end{cases} \tag{6.71}$$

$A_c(y, x_c)$ is the input function of the integrator.

Problem formulation: Given any $\gamma > 0$, find a controller of the form (6.71) such that the L_2 gain from the disturbance $\omega(k)$ to the output that needs be controlled by $z(k)$ for system (6.68) with (6.71) is less than or equal to γ, i.e.,

$$\|z(k)\|_{[0,\infty]} \leq \gamma^2 \|\omega(k)\|_{[0,\infty]} \tag{6.72}$$

for all $w(k) \in L_2[0, \infty]$ and all admissible uncertainties. In this situation, system (6.68) is said to have a robust H_∞ performance (6.72).

Motivated by [34], we define the following "*scaled*" system:

$$\begin{cases} \tilde{x}(k+1) = A(\tilde{x}(k))\tilde{x}(k) + \begin{bmatrix} B_\omega(\tilde{x}(k)) & \frac{1}{\delta}\bar{H}(\tilde{x}(k)) \end{bmatrix} \tilde{\omega}(k) + B_u(\tilde{x}(k))u(k), \\ \tilde{z}(k) = \begin{bmatrix} C_z(\tilde{x}(k)) \\ \delta E_1(\tilde{x}(k)) \end{bmatrix} \tilde{x}(k) + \begin{bmatrix} D_{zu}(\tilde{x}(k)) \\ \delta E_2(\tilde{x}(k)) \end{bmatrix} u(k), \\ \tilde{y}(k) = C_y(x(k))x(k), \end{cases} \tag{6.73}$$

where $\tilde{x} \in R^n$ is the state, $u(k) \in R^m$, $\tilde{\omega} \in R^{m+i}$ is the input noise, δ is a positive constant, $\tilde{z}(k)$ is the controlled output, and $\bar{H}(\tilde{x}(k)) = [H_1(\tilde{x}(k)) \quad H_1(\tilde{x}(k))]$.

Remark 30. System (6.73) is similar to system (6.1). Therefore, we can apply the methodology used for solving system (6.1) for the "*scaled*" system (6.73).

System (6.73) with controller (6.71) can be written as follows:

$$\begin{cases} \hat{x}(k+1) = \hat{A}(\hat{x}(k))\hat{x}(k) + \hat{B}_u(\hat{x}(k))A_c(y, x_c) + \hat{B}_\omega(\hat{x}(k))\tilde{\omega}(k), \\ \tilde{z}(k) = \hat{C}_z(\hat{x}(k))\hat{x}(k), \\ \tilde{y}(k) = \hat{C}_y(\hat{x}(k))\hat{x}(k), \end{cases} \tag{6.74}$$

where

$$\hat{A}(\hat{x}(k)) = \begin{bmatrix} A(\tilde{x}(k)) & B_u(\tilde{x}(k)) \\ 0 & 1 \end{bmatrix}, \quad \hat{B}_u(\hat{x}(k)) = \begin{bmatrix} 0 \\ 1 \end{bmatrix},$$
$$\hat{B}_\omega(\hat{x}(k)) = \begin{bmatrix} \tilde{B}_\omega(\tilde{x}(k)) \\ 0 \end{bmatrix}, \quad \hat{C}_z(\hat{x}(k)) = \begin{bmatrix} \tilde{C}_z(\tilde{x}(k)) & \tilde{D}_{zu}(\tilde{x}(k)) \end{bmatrix},$$
$$\hat{C}_y(\hat{x}(k)) = \begin{bmatrix} C_y(x(k)) & 0 \end{bmatrix}, \quad \hat{x} = \begin{bmatrix} x(k) \\ x_c(k) \end{bmatrix}, \tag{6.75}$$

with

$$\tilde{B}_\omega(\tilde{x}(k)) = \begin{bmatrix} B_\omega(\tilde{x}(k)) & \frac{1}{\delta}\bar{H}(\tilde{x}(k)) \end{bmatrix}, \quad \tilde{C}_z(\tilde{x}(k)) = \begin{bmatrix} C_z(\tilde{x}(k)) \\ \delta E_1(\tilde{x}(k)) \end{bmatrix},$$
$$\tilde{D}_{zu}(\tilde{x}(k)) = \begin{bmatrix} D_{zu}(\tilde{x}(k)) \\ \delta E_2(\tilde{x}(k)) \end{bmatrix}. \tag{6.76}$$

Next, we assume $A_c(y, x_c)$ to be of the form $A_c(y, x_c) = \hat{A}_c(y, x_c)y$. Therefore, (6.74) can be rewritten as follows:

$$\begin{cases} \hat{x}(k+1) = \hat{A}(\hat{x}(k))\hat{x}(k) + \hat{B}_u(\hat{x}(k))\hat{A}_c(y, x_c)y + \hat{B}_\omega(\hat{x}(k))\tilde{\omega}(k), \\ \tilde{z}(k) = \hat{C}_z(\hat{x}(k))\hat{x}(k), \\ \tilde{y}(k) = \hat{C}_y(\hat{x}(k))\hat{x}(k), \end{cases} \tag{6.77}$$

where $\hat{A}(\hat{x}(k))$, $\hat{B}_u(\hat{x}(k))$, $\hat{B}_\omega(\hat{x}(k))$, $\hat{C}_z(\hat{x}(k))$, and $\hat{C}_y(\hat{x}(k))$ are as described in (6.75).

In view of the "*scaled*" system (6.73), we establish the following theorem.

Theorem 17. *Consider system* (6.68). *There exists a controller of the form* (6.71) *such that* (6.72) *holds for all admissible uncertainties if there exists* *a positive constant $\delta > 0$ such that (6.72) holds for system (6.73) with the same controller.*

Proof. The proof can be shown using similar techniques as proposed in [34] and hence omitted here. □

In the light of Theorem 17, it remains to solve the "*scaled*" nonlinear H_∞ control problem given in (6.73). Therefore, sufficient conditions for the existence of a solution to the robust H_∞ output feedback control problem are presented in the following theorem.

Theorem 18. *Given a prescribed H_∞ performance $\gamma > 0$, system (6.68) is stabilizable with H_∞ performance (6.72) via the nonlinear output feedback controller of the form (6.71) if there exist a symmetric polynomial matrix $\hat{P}(x(k))$, a polynomial function $L_{31}(y, x_c)$, and a polynomial matrix $\hat{G}(\hat{x}(k))$ such that the following conditions hold for all $x \neq 0$:*

$$\hat{P}(x(k)) > 0, \tag{6.78}$$

$$M_2(\hat{x}(k)) > 0, \tag{6.79}$$

where

$$M_2(\hat{x}(k)) = \begin{bmatrix} \hat{P}(x(k)) & * & * & * \\ 0 & \gamma^2 I & * & * \\ \hat{G}(\hat{x}(k))\hat{A}(\hat{x}(k)) + \hat{L}(y, x_c)\hat{C}_y(\hat{x}(k)) & \hat{B}_\omega(\hat{x}(k)) & \hat{G}(\hat{x}(k)) + \hat{G}^T(\hat{x}(k)) - \hat{P}(x_+) & * \\ \hat{C}_z(\hat{x}(k)) & 0 & 0 & I \end{bmatrix}, \tag{6.80}$$

with

$$\hat{G}(\hat{x}(k)) = \begin{bmatrix} G_{11}(\hat{x}(k)) & G_{12}(\hat{x}(k)) & 0 \\ G_{21}(\hat{x}(k)) & G_{22}(\hat{x}(k)) & 0 \\ G_{31}(\hat{x}(k)) & G_{32}(\hat{x}(k)) & G_{33}(y, x_c) \end{bmatrix},$$

$$\hat{L}(y, x_c) = \begin{bmatrix} 0 \\ 0 \\ L_{31}(y, x_c) \end{bmatrix}. \tag{6.81}$$

Moreover, the nonlinear output feedback controller is given by

$$\begin{aligned} x_c(k+1) &= x_c(k) + A_c(y, x_c), \\ u(k) &= x_c(k), \end{aligned} \tag{6.82}$$

where

$$A_c(y, x_c) = \hat{K}(y, x_c)\hat{C}_y(\hat{x}(k))\hat{x}(k)$$
$$\text{with} \quad \hat{K}(y, x_c) = L_{31}(y, x_c)G_{33}^{-1}(y, x_c). \tag{6.83}$$

Proof. By Theorem 17 the robust nonlinear H_∞ output feedback control problem is converted to the nonlinear H_∞ output feedback control problem for a "*scaled*" system. Then, by adapting Theorem 16 the result can be obtained easily. It is important to note that a Lyapunov function of the following form is selected:

$$\hat{V}(\hat{x}(k)) = \hat{x}^T(k)\hat{P}^{-1}(x(k))\hat{x}(k). \tag{6.84}$$

□

Remark 31. The idea of choosing a Lyapunov function to be of the form (6.84) is to ensure a convex solution to the terms in $P(x(k+1))$ can be achieved.

Note that conditions (6.78)–(6.79) of Theorem 18 are in state-dependent polynomial matrix inequalities (PMIs). Using the SOS decomposition method based on SDP [31] provides a relaxation for the problem. Therefore, (6.78)–(6.79) can be modified into SOS conditions given in the following corollary.

Corollary 11. *Given a prescribed H_∞ performance $\gamma > 0$, system* (6.68) *is asymptotically stable via the nonlinear output feedback controller* (6.71) *with H_∞ performance* (6.72) *if there exist a symmetric polynomial matrix $\hat{P}(x(k))$, a polynomial function $L_{31}(y, x_c)$, a polynomial matrix $\hat{G}(\hat{x}(k))$, and constants $\epsilon_1 > 0$ and $\epsilon_2 > 0$ such that the following conditions hold for all $x \neq 0$:*

$$v_5^T[\hat{P}(x(k)) - \epsilon_1 I]v_5 \quad \text{is an SOS}, \tag{6.85}$$
$$v_6^T\left[M_2(\hat{x}(k)) - \epsilon_2 I\right]v_6 \quad \text{is an SOS}, \tag{6.86}$$

where v_5 and v_6 are vectors of appropriate dimensions, and

$$M_2(\hat{x}(k)) =$$
$$\begin{bmatrix} \hat{P}(x(k)) & * & * & * \\ 0 & \gamma^2 I & * & * \\ \hat{G}(\hat{x}(k))\hat{A}(\hat{x}(k)) + \hat{L}(y, x_c)\hat{C}_y(\hat{x}(k)) & \hat{B}_\omega(\hat{x}(k)) & \hat{G}(\hat{x}(k)) + \hat{G}^T(\hat{x}(k)) - \hat{P}(x_+) & * \\ \hat{C}_z(\hat{x}(k)) & 0 & 0 & I \end{bmatrix} \tag{6.87}$$

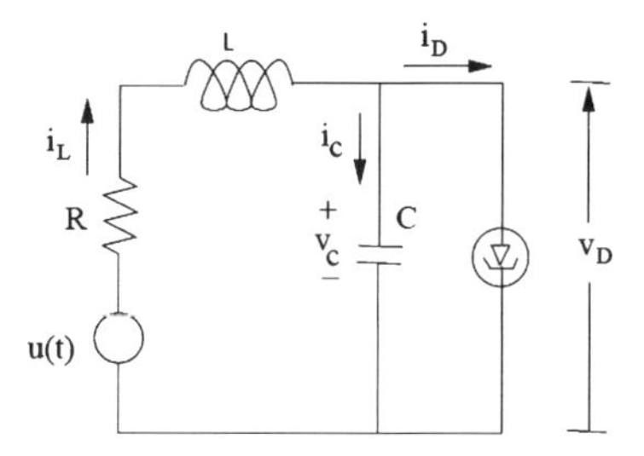

■ **FIGURE 6.1** Tunnel diode circuit.

with

$$\hat{G}(\hat{x}(k)) = \begin{bmatrix} G_{11}(\hat{x}(k)) & G_{12}(\hat{x}(k)) & 0 \\ G_{21}(\hat{x}(k)) & G_{22}(\hat{x}(k)) & 0 \\ G_{31}(\hat{x}(k)) & G_{32}(\hat{x}(k)) & G_{33}(y, x_c) \end{bmatrix},$$

$$\hat{L}(y, x_c) = \begin{bmatrix} 0 \\ 0 \\ L_{31}(y, x_c) \end{bmatrix}. \tag{6.88}$$

Moreover, the nonlinear output feedback controller is given by

$$\begin{aligned} x_c(k+1) &= x_c(k) + A_c(y, x_c), \\ u(k) &= x_c(k), \end{aligned} \tag{6.89}$$

where

$$\begin{aligned} &A_c(y, x_c) = \hat{K}(y, x_c)\hat{C}_y(\hat{x}(k))\hat{x}(k) \\ &\quad \textit{with} \quad \hat{K}(y, x_c) = L_{31}(y, x_c)G_{33}^{-1}(y, x_c). \end{aligned} \tag{6.90}$$

Proof. The proof for this section follows directly by combining the proofs of Theorem 18 and Lemma 3. The proof ends. □

6.4 NUMERICAL EXAMPLES

Example 1. **A Tunnel Diode Circuit.** A tunnel diode circuit with input $u(t)$ is shown in Fig. 6.1 [35]. The characteristics of the tunnel diode are described as follows:

$$i_D(t) = 0.002v_D(t) + 0.01v_D^3(t). \tag{6.91}$$

Next, choosing the state variables of the form $x_1(t) = v_c(t)$ and

$x_2(t) = i_L(t)$, the circuit can be represented by the following state equations:

$$\begin{aligned} C\dot{x}_1(t) &= -0.002x_1(t) - 0.01x_1^3(t) + x_2(t), \\ L\dot{x}_2(t) &= -x_1(t) - Rx_2(t) + \omega(t) + u(t), \\ y(t) &= Sx(t), \\ z(t) &= x_2(t) + u(t), \end{aligned} \tag{6.92}$$

where $\omega(t)$ is the noise to the system, $y(t)$ is the measured output, $z(t)$ is the controlled output, and $u(t)$ is the input to the circuit. In addition, we assume that the state $x_2(t) = i_L(t)$ is available for feedback. Therefore, $S = [0 \quad 1]$. The circuit parameter is given as follows: $C = 20$ mF, $\epsilon = 1000$ mH, and $R = 1\ \Omega$. With these parameters, the dynamic of the circuit can be written as follows:

$$\begin{aligned} \dot{x}_1(t) &= -0.1x_1(t) - 0.5x_1^3(t) + 50x_2(t), \\ \dot{x}_2(t) &= -x_1(t) - x_2(t) + \omega(t) + u(t), \\ y(t) &= x_2(t), \\ z(t) &= x_2(t) + u(t). \end{aligned} \tag{6.93}$$

The above system is in continuous time; therefore, to convert (6.93) into discrete time, we sample the above system at $T = 0.02$, and applying Euler's discretization method, we obtain the following discrete-time nonlinear dynamic equations:

$$\begin{aligned} x_1(k+1) &= x_1(k) + T\big[-0.1x_1(k) - 0.5x_1^3(k) + 50x_2(k)\big], \\ x_2(k+1) &= x_2(k) + T\big[-x_1(t) - x_2(t) + \omega(t) + u(t)\big], \\ y(k) &= x_2(k), \\ z(k) &= x_2(k) + u(k). \end{aligned} \tag{6.94}$$

Then, from (6.94), the system with controller (6.16) can be written as follows:

$$\begin{cases} \hat{x}(k+1) = \hat{A}(\hat{x}(k))\hat{x}(k) + \hat{B}_u(\hat{x}(k))\hat{A}_c(\hat{x}(k))\hat{x}(k) + \hat{B}_\omega(\hat{x}(k))\omega(k), \\ z(k) = \hat{C}_z(\hat{x}(k))\hat{x}(k), \\ y(k) = \hat{C}_y(\hat{x}(k))\hat{x}(k), \end{cases} \tag{6.95}$$

where

$$\hat{A}(\hat{x}(k)) = \begin{bmatrix} 1 + T\big(-0.1 - 0.5x_1^2(k)\big) & 50T & 0 \\ -T & 1-T & T \\ 0 & 0 & 1 \end{bmatrix},$$

$$\hat{B}_u(\hat{x}(k)) = \begin{bmatrix} 0 \\ 0 \\ 1 \end{bmatrix}, \quad \hat{B}_\omega(\hat{x}(k)) = \begin{bmatrix} 0 \\ T \\ 0 \end{bmatrix}, \quad \hat{C}_z(\hat{x}(k)) = \begin{bmatrix} 0 & 1 & 1 \end{bmatrix},$$

$$\hat{C}_y(\hat{x}(k)) = \begin{bmatrix} 0 & 1 & 0 \end{bmatrix}, \quad \hat{x}(k) = \begin{bmatrix} x_1(k) \\ x_2(k) \\ x_c(k) \end{bmatrix}. \tag{6.96}$$

In this example, we choose $\epsilon_1 = \epsilon_2 = 0.01$ and $\gamma = 1$. Then, using the procedure described in Corollary 9 and with $\hat{P}(x(k))$ and $\hat{G}(\hat{x}(k))$ of degree 4 and $\hat{L}(y, x_c)$ of degree 8, a feasible solution is achieved. The SOSTOOLS returns the following values:

$$\begin{aligned} G_{33}(\hat{x}(k)) &= 14.68x_2^4 - 0.085x_2^3x_c + 0.0102x_2^2x_c^2 - 59.401x_2^2 \\ &\quad - 0.004x_2x_c^3 + 1.828x_2x_c + 0.007x_c^4 - 0.484x_c^2 + 7858.733, \\ L_{31}(\hat{x}(k)) &= -1.009x_2^4 - 0.403x_2^3x_c - 0.0026x_2^2x_c^2 - 0.00436x_2x_c^3 \\ &\quad + 8.705x_2x_c - 0.00033x_c^4 + 0.113x_c^2 - 285.65. \end{aligned} \tag{6.97}$$

The nonlinear controller is given by

$$\begin{aligned} x_c(k+1) &= x_c(k) + K(y, x_c)C(\hat{x}(k))\hat{x}(k), \\ u(k) &= x_c(k), \end{aligned} \tag{6.98}$$

where

$$K(y, x_c) = \frac{L_{31}(y, x_c)}{G_{33}(\hat{x}(k))}. \tag{6.99}$$

It is best to note that calculation of $K(y, x_c)$ is hard because it contains the polynomial terms; hence, we use Matlab/Simulink to compute those values. It is also important to highlight here that the Lyapunov matrix $\hat{P}(x(k))$ in this example is defined to be a symmetrical $N \times N$ polynomial matrices whose (i, j)th entry is given by

$$p_{ij}(x(k)) = p_{ij}^0 + p_{ijg}m(k)^{(1:l)}, \tag{6.100}$$

where $i = 1, 2, \ldots, n$, $j = 1, 2, \ldots, n$, and $g = 1, 2, \ldots, d$, where n is the number of states, and d is the total number of monomials. Meanwhile, $m(k)$ is the number of all monomial vectors in $(x(k))$ from degree of 1 to degree of l, where l is a scalar even value. For example, if $l = 2$ and $x(k) = [x_1(k), x_2(k)]^T$, then $p_{11}(x(k)) = p_{11} + p_{112}x_1 + p_{113}x_2 + p_{114}x_1^2 + p_{115}x_1x_2 + p_{116}x_2^2$. This structure is more general as compared to [30]; because of a higher value of l, a more relaxation in SOS problem can be achieved. Due to the large number of values returned by SOSTOOLS for $P(x(k))$ in this example, those values are omitted.

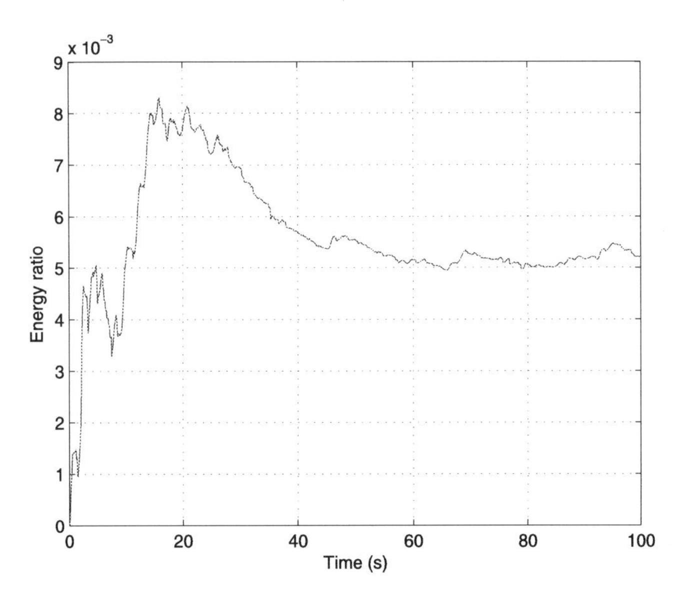

■ **FIGURE 6.2** Ratio of the regulated output energy to the disturbance input noise energy of a tunnel diode circuit.

Remark 32. The disturbance input signal $w(k)$ used during the simulation is the band-limited white noise (noise power is 10). The simulation result for the ratio of the controlled output energy to the disturbance input noise energy obtained by using the H_∞ output feedback controller is illustrated in Fig. 6.2. It can be seen from the figure that the ratio of the controlled output energy to the disturbance input noise energy is always less than a prescribed value 1 and decreases to about 0.005. Thus, $\gamma = \sqrt{0.005} \approx 0.07$. This implies that the L_2 gain from the disturbance to the regulated output is no greater than 0.07.

Remark 33. To date, to the authors' knowledge, no result has been presented in the framework of H_∞ output feedback control for polynomial discrete-time systems.

Example 2. **With parametric uncertainties.** Consider the tunnel diode circuit shown in Fig. 6.1. For this example, we assume that the value of R is uncertain and given by $R = 1 \pm 30\%\ \Omega$. Therefore, the system can be described as follows:

$$\hat{A}(\hat{x}(k)) = \begin{bmatrix} 1 + T\big(-0.1 - 0.5x_1^2(k)\big) & 50T & 0 \\ -T & 1 - RT & T \\ 0 & 0 & 1 \end{bmatrix},$$

$$\hat{B}_u(\hat{x}(k)) = \begin{bmatrix} 0 \\ 0 \\ 1 \end{bmatrix}, \quad \hat{B}_\omega(\hat{x}(k)) = \begin{bmatrix} 0 \\ T \\ 0 \end{bmatrix}, \quad \hat{C}_z(\hat{x}(k)) = \begin{bmatrix} 0 & 1 & 1 \end{bmatrix},$$
$$\hat{C}_y(\hat{x}(k)) = \begin{bmatrix} 0 & 1 & 0 \end{bmatrix}. \tag{6.101}$$

For this example, we choose $\epsilon_1 = \epsilon_2 = 0.01$ and $\gamma = 1$. Then, using the procedure described in Corollary 10 with $\hat{P}(x(k))$ and $\hat{G}(\hat{x}(k))$ of degree 4 and $\hat{L}(y, x_c)$ of degree 6, a feasible solution is achieved. The following values are returned by SOSTOOLS:

$$\begin{aligned} G_{33}(\hat{x}(k)) &= 14.18x_2^4 - 0.076x_2^3x_c + 0.0062x_2^2x_c - 11.867x_2^2 \\ &\quad - 0.0119x_2x_c^3 + 0.463x_2x_c + 0.0022x_c^4 - 0.089x_c^2 + 328.0373, \\ L_{31}(\hat{x}(k)) &= -4.591x_2^4 + 1.169x_2^3x_c - 0.9036x_2^2x_c^2 + 0.007x_2x_c^3 \\ &\quad + 2.0713x_2x_c + 0.1765x_c^4 - 15.027x_2^2 + 0.455x_c^2 - 13.614. \end{aligned} \tag{6.102}$$

The nonlinear controller is given by

$$\begin{aligned} x_c(k+1) &= x_c(k) + K(y, x_c)C(\hat{x}(k))\hat{x}(k), \\ u(k) &= x_c(k), \end{aligned} \tag{6.103}$$

where

$$K(y, x_c) = \frac{L_{31}(y, x_c)}{G_{33}(\hat{x}(k))}. \tag{6.104}$$

The disturbance input signal $w(k)$ used during the simulation is the band-limited white noise (noise power is 2). The simulation results for the ratio of the controlled output energy to the disturbance input noise energy obtained by using the H_∞ output feedback controller is shown in Fig. 6.3. It can be seen from the figure that the ratio of the controlled output energy to the disturbance input noise energy is always less than the prescribed value 1 and decreasing to about 0.00012. Thus, $\gamma = \sqrt{0.00012} \approx 0.011$. This implies that the L_2 gain from the disturbance to the regulated output is no greater than 0.011.

Example 3. **Norm-Bounded Uncertainties.** A tunnel diode circuit with input $u(t)$ is shown in Fig. 6.1 [35]. The characteristics of the tunnel diode are described as follows:

$$i_D(t) = 0.002v_D(t) + 0.01v_D^3(t). \tag{6.105}$$

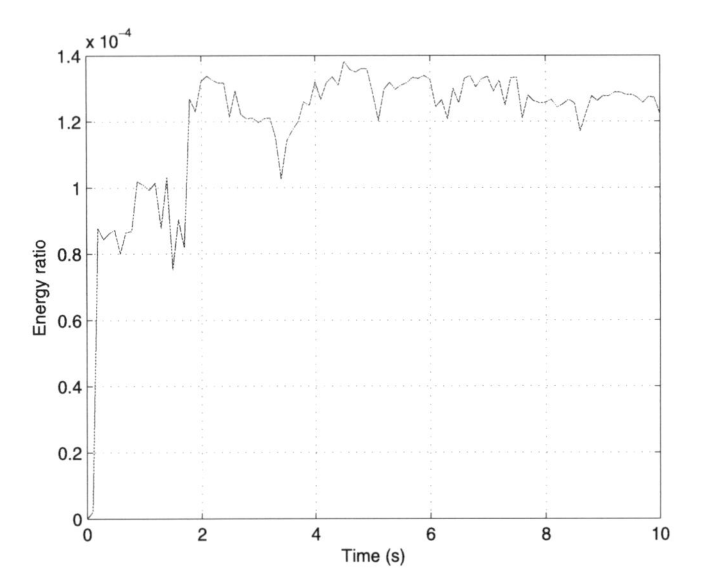

■ **FIGURE 6.3** Ratio of the regulated output energy to the disturbance input noise energy of a tunnel diode circuit with polytopic uncertainty.

Next, choosing the state variables of the form $x_1(t) = v_c(t)$ and $x_2(t) = i_L(t)$, the circuit can be represented by the following state equations:

$$\begin{aligned} C\dot{x}_1(t) &= -0.002x_1(t) - 0.01x_1^3(t) + x_2(t), \\ L\dot{x}_2(t) &= -x_1(t) - Rx_2(t) + \omega(t) + u(t), \\ y(t) &= Sx(t), \\ z(t) &= x_2(t) + u(t), \end{aligned} \tag{6.106}$$

where $\omega(t)$ is the noise to the system, $y(t)$ is the measured output, $z(t)$ is the controlled output, and $u(t)$ is the input to the circuit. In addition, we assume that the state $x_2(t) = i_L(t)$ is available for feedback. Therefore, $S = [0 \quad 1]$. The circuit parameter is given as follows: $C = 20$ mF, $\epsilon = 1000$ mH, and $R = 1 \pm 30\%$ Ω. With these parameters, the dynamics of the circuit can be written as follows:

$$\begin{aligned} \dot{x}_1(t) &= -0.1x_1(t) - 0.5x_1^3(t) + 50x_2(t), \\ \dot{x}_2(t) &= -x_1(t) - Rx_2(t) + \omega(t) + u(t), \\ y(t) &= x_2(t), \\ z(t) &= x_2(t) + u(t). \end{aligned} \tag{6.107}$$

The above system is in continuous time; therefore, to convert (6.107) into discrete time, we sample the above system at $T = 0.02$, and by Euler's discretization method, the following discrete-time nonlinear dynamic equations are obtained:

$$\begin{aligned} x_1(k+1) &= x_1(k) + T\big[-0.1x_1(k) - 0.5x_1^3(k) + 50x_2(k)\big], \\ x_2(k+1) &= x_2(k) + T\big[-x_1(t) - Rx_2(t) + \omega(t) + u(t)\big], \\ y(k) &= x_2(k) \\ z(k) &= x_2(k) + u(k). \end{aligned} \tag{6.108}$$

From (6.108), the system with controller (6.71) can be written as follows:

$$\begin{cases} \hat{x}(k+1) = \hat{A}(\hat{x}(k))\hat{x}(k) + \hat{B}_u(\hat{x}(k))\hat{A}_c(\hat{x}(k))\hat{x}(k) + \hat{B}_\omega(\hat{x}(k))\omega(k), \\ z(k) = \hat{C}_z(\hat{x}(k))\hat{x}(k), \\ y(k) = \hat{C}_y(\hat{x}(k))\hat{x}(k), \end{cases} \tag{6.109}$$

where

$$\begin{aligned} &\hat{A}(\hat{x}(k)) = \begin{bmatrix} A(\tilde{x}(k)) & B_u(\tilde{x}(k)) \\ 0 & 1 \end{bmatrix}, \quad \hat{B}_u(\hat{x}(k)) = \begin{bmatrix} 0 \\ 1 \end{bmatrix}, \\ &\hat{B}_\omega(\hat{x}(k)) = \begin{bmatrix} \tilde{B}_\omega(\tilde{x}(k)) \\ 0 \end{bmatrix}, \quad \hat{C}_z(\hat{x}(k)) = \begin{bmatrix} \tilde{C}_z(\tilde{x}(k)) & \tilde{D}_{zu}(\tilde{x}(k)) \end{bmatrix}, \\ &\hat{C}_y(\hat{x}(k)) = \begin{bmatrix} C_y(x(k)) & 0 \end{bmatrix}, \quad \hat{x} = \begin{bmatrix} x(k) \\ x_c(k) \end{bmatrix}, \end{aligned} \tag{6.110}$$

with

$$\begin{aligned} &\tilde{B}_\omega(\tilde{x}(k)) = \begin{bmatrix} B_\omega(\tilde{x}(k)) & \frac{1}{\delta}\bar{H}(\tilde{x}(k)) \end{bmatrix}, \quad \tilde{C}_z(\tilde{x}(k)) = \begin{bmatrix} C_z(\tilde{x}(k)) \\ \delta E_1(\tilde{x}(k)) \end{bmatrix}, \\ &\tilde{D}_{zu}(\tilde{x}(k)) = \begin{bmatrix} D_{zu}(\tilde{x}(k)) \\ \delta E_2(\tilde{x}(k)) \end{bmatrix}. \end{aligned} \tag{6.111}$$

Similarly,

$$\hat{A}(\hat{x}(k)) = \begin{bmatrix} 1 + T\big[-0.1 - 0.5x_1^2(k)\big] & 50T & 0 \\ -T & 1 - RT & T \\ 0 & 0 & 1 \end{bmatrix},$$

$$\hat{B}_u(\hat{x}(k)) = \begin{bmatrix} 0 \\ 0 \\ 1 \end{bmatrix}, \quad \hat{B}_\omega(\hat{x}(k)) = \begin{bmatrix} 0 & 0 & 0 \\ T & \frac{1}{\delta}0.3 & \frac{1}{\delta}0.3 \\ 0 & 0 & 0 \end{bmatrix},$$

$$\hat{C}_z(\hat{x}(k)) = \begin{bmatrix} 0 & 1 & 1 \\ 0 & \delta & 0 \end{bmatrix}, \quad \hat{C}_y(\hat{x}(k)) = \begin{bmatrix} 0 & 1 & 0 \end{bmatrix}, \quad \hat{x}(k) = \begin{bmatrix} x_1(k) \\ x_2(k) \\ x_c(k) \end{bmatrix}, \tag{6.112}$$

where $\delta = 1$.

In this example, we choose $\epsilon_1 = \epsilon_2 = 0.01$ and $\gamma = 1$. Then, using the procedure described in the Corollary 11 with $\hat{P}(x(k))$ and $\hat{G}(\hat{x}(k))$ of degree 4 and $\hat{L}(y, x_c)$ of degree 8, a feasible solution is achieved. The following values are returned by SOSTOOLS:

$$\begin{aligned} G_{33}(\hat{x}(k)) &= 25.5662x_2^4 - 0.0351x_2^3x_c + 0.1624x_2^2x_c^2 - 38.6407x_2^2 \\ &\quad - 0.0044x_2x_c^3 + 0.0213x_2x_c + 0.01174x_c^4 - 0.5731x_c^2 \\ &\quad + 6162.2353, \\ L_{31}(\hat{x}(k)) &= -0.0002x_2^6 + 0.0002x_2^4x_c^2 - 3.0236x_2^4 - 0.0003x_2^2x_c^4 \\ &\quad - 0.0128x_2^2x_c^2 + 11.7213x_2^2 - 0.0112x_2x_c^3 + 12.4267x_2x_c \\ &\quad - 0.0007x_c^4 + 0.1586x_c^2 - 265.2317. \end{aligned} \tag{6.113}$$

The nonlinear controller is given by

$$\begin{aligned} x_c(k+1) &= x_c(k) + K(y, x_c)C(\hat{x}(k))\hat{x}(k), \\ u(k) &= x_c(k), \end{aligned} \tag{6.114}$$

where

$$K(y, x_c) = \frac{L_{31}(y, x_c)}{G_{33}(\hat{x}(k))}. \tag{6.115}$$

It is best to not that calculation of $K(y, x_c)$ is hard because it contains the polynomial terms; hence, we use Matlab/Simulink to compute those values. It is also important to highlight here that the Lyapunov matrix $\hat{P}(x(k))$ in this example is defined to be a symmetrical $N \times N$ polynomial matrix whose (i, j)th entry is described in (6.100).

Remark 34. The disturbance input signal $w(k)$ used during the simulation is the band-limited white noise (noise power is 1). The simulation results for the ratio of the controlled output energy to the disturbance input noise energy is illustrated in Fig. 6.4. It can be seen from the figure that the ratio of the controlled output energy to the disturbance input noise energy is always less than a prescribed value 1 and decreasing to about 0.007. Thus, $\gamma = \sqrt{0.007} \approx 0.083$. This implies that the L_2 gain from the disturbance to the regulated output is no greater than 0.08.

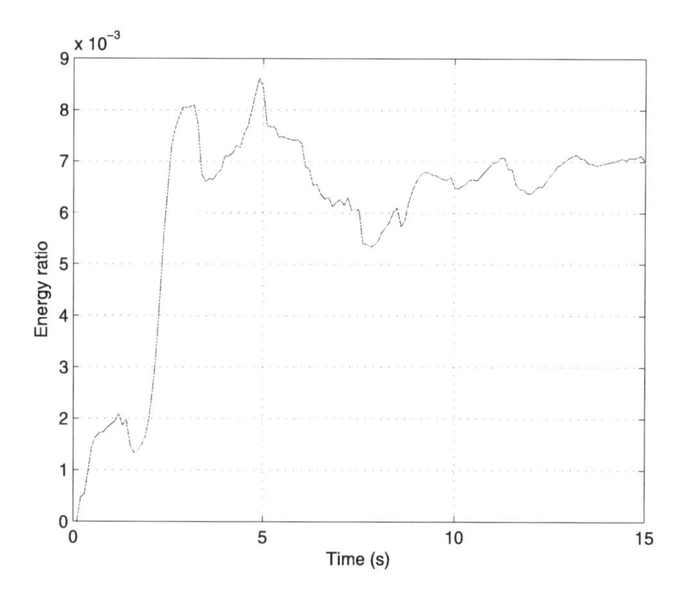

■ **FIGURE 6.4** Ratio of the regulated output energy to the disturbance input noise energy of a tunnel diode circuit.

6.5 CONCLUSION

The problem of designing a robust nonlinear H_∞ output feedback controller for polynomial discrete-time systems has been examined in this chapter. In particular, both parametric and norm-bounded uncertainty have been considered in this chapter. The interconnection between the robust nonlinear H_∞ output feedback control problem and nonlinear H_∞ output feedback control problem has been established through a so-called "*scaled*" system. The integrator is incorporated into the controller structures, and hence a convex solution in $P(x(k+1))$ can be obtained efficiently. Sufficient conditions for the existence of the proposed controller are provided in terms of the solvability of PMIs, which are formulated as SOS constraints and have been solved using SOSTOOLS. The effectiveness of the proposed design methodology is demonstrated through a tunnel diode circuit.

REFERENCES

[1] V.L. Syrmos, C.T. Abdallah, P. Dorato, K. Grigoriadis, Static output feedback: a survey, Automatica 33 (1997) 125–137.

[2] F. Rasool, D. Huang, S.K. Nguang, Robust H_∞ output feedback control of networked control systems with multiple quantizers, Journal of the Franklin Institute 349 (3) (2012) 1153–1173.

[3] F. Rasool, D. Huang, S.K. Nguang, Robust H_∞ output feedback control of discrete-time networked systems with limited information, Systems & Control Letters 60 (10) (2011) 845–853.
[4] S.K. Nguang, W. Assawinchaichote, P. Shi, Y. Shi, Robust H_∞ control design for uncertain fuzzy systems with Markovian jumps: an LMI approach, in: American Control Conference, 2005, pp. 1805–1810.
[5] S.K. Nguang, P. Zhang, S.X. Ding, Parity relation based fault estimation for nonlinear systems: an LMI approach, International Journal of Automation and Computing 4 (2) (2007) 164–168.
[6] S.K. Nguang, Comments on "Robust stabilization of uncertain input-delay systems by sliding mode control with delay compensation", Automatica 37 (10) (2001) 1677.
[7] S.K. Nguang, P. Shi, H_∞ output feedback control of fuzzy system models under sampled measurements, Computers and Mathematics with Applications 46 (5) (2003) 705–717.
[8] W. Assawinchaichote, S.K. Nguang, P. Shi, Fuzzy Control and Filter Design for Uncertain Fuzzy Systems, Springer, 2006.
[9] S.K. Nguang, P. Shi, On designing filters for uncertain sampled-data nonlinear systems, Systems & Control Letters 41 (5) (2000) 305–316.
[10] D. Huang, S.K. Nguang, Robust Control for Uncertain Networked Control Systems With Random Delays, Springer Science & Business Media, 2009.
[11] J. Zhang, A.K. Swain, S.K. Nguang, Robust sensor fault estimation scheme for satellite attitude control systems, Journal of the Franklin Institute 350 (9) (2013) 2581–2604.
[12] J. Zhang, P. Shi, J. Qiu, S.K. Nguang, A novel observer-based output feedback controller design for discrete-time fuzzy systems, IEEE Transactions on Fuzzy Systems 23 (1) (2015) 223–229.
[13] S.K. Nguang, P. Shi, Delay-dependent H_∞ filtering for uncertain time delay nonlinear systems: an LMI approach, IET Control Theory & Applications 1 (1) (2007) 133–140.
[14] Z. Hou, J. Luo, P. Shi, S.K. Nguang, Stochastic stability of Ito differential equations with semi-Markovian jump parameters, IEEE Transactions on Automatic Control 51 (8) (2006) 1383–1387.
[15] W. Assawinchaichote, S.K. Nguang, P. Shi, E.K. Boukas, H_∞ fuzzy state-feedback control design for nonlinear systems with-stability constraints: an LMI approach, Mathematics and Computers in Simulation 78 (4) (2008) 514–531.
[16] S. Chae, S.K. Nguang, SOS based robust H_∞ fuzzy dynamic output feedback control of nonlinear networked control systems, IEEE Transactions on Cybernetics 44 (7) (2014) 1204–1213.
[17] S.K. Nguang, P. Shi, Stabilisation of a class of nonlinear time-delay systems using fuzzy models, in: Proceedings of the 39th IEEE Conference on Decision and Control, 2000, pp. 5–11.
[18] S. Saat, S.K. Nguang, Nonlinear H_∞ output feedback control with integrator for polynomial discrete-time systems, International Journal of Robust and Nonlinear Control 25 (2015) 1051–1065.
[19] Y. Zhang, P. Shi, S.K. Nguang, H.R. Karimi, Observer-based finite-time fuzzy H_∞ control for discrete-time systems with stochastic jumps and time-delays, Signal Processing 97 (2014) 252–261.

[20] S. Chae, F. Rasool, S.K. Nguang, A. Swain, Robust mode delay-dependent H_∞ control of discrete-time systems with random communication delays, IET Control Theory & Applications 4 (6) (2010) 936–944.

[21] Y. Zhang, P. Shi, S.K. Nguang, Y. Song, Robust finite-time H_∞ control for uncertain discrete-time singular systems with Markovian jumps, IET Control Theory & Applications 8 (12) (2014) 1105–1111.

[22] D. Zhao, J. Wang, Robust static output feedback design for polynomial nonlinear systems, International Journal of Robust and Nonlinear Control (2009).

[23] S. Saat, M. Krug, S.K. Nguang, A nonlinear static output controller design for polynomial systems: an iterative sums of squares approach, in: 4th International Conference on Control and Mechatronics (ICOM), 2011, pp. 1–6.

[24] S. Saat, M. Krug, S.K. Nguang, Nonlinear H_∞ static output feedback controller design for polynomial systems: an iterative sums of squares approach, in: 6th IEEE Conference on Industrial Electronics and Applications (ICIEA), 2011, pp. 985–990.

[25] Jiuxiang Dong, G.H. Yang, Static output feedback control synthesis for discrete-time T-S fuzzy systems, International Journal of Control, Automation, and Systems 5 (2007) 349–1354.

[26] Jiuxiang Dong, G.H. Yang, Static output feedback H_∞ control of a class of nonlinear discrete-time systems, Fuzzy Sets and Systems 160 (2009) 2844–2859.

[27] Q. Zheng, F. Wu, Regional stabilisation of polynomial nonlinear systems using rational Lyapunov functions, International Journal of Control 82 (9) (2009) 1605–1615.

[28] M.C. de Oliveira, J. Bernussou, J.C. Geromel, A new discrete-time robust stability condition, Systems & Control Letters 37 (4) (1999) 261–265.

[29] S. Prajna, A. Papachristodoulou, P.A. Parrilo, Introducing SOSTOOLS: a general purpose sum of squares programming solver, in: Conference on Decision and Control, vol. 1, 2002, pp. 741–746.

[30] H.J. Ma, G.H. Yang, Fault tolerant H_∞ control for a class of nonlinear discrete-time systems: using sum of squares optimization, in: Proceedings of American Control Conference, 2008, pp. 1588–1593.

[31] S. Prajna, A. Papachristodoulou, F. Wu, Nonlinear control synthesis by sum of squares optimization: a Lyapunov-based approach, in: Proceedings of the 5th Asian Control Conference, 2004, pp. 157–165.

[32] P.A. Parrilo, Structured Semidefinite Programs and Semialgebraic Geometry Methods in Robustness and Optimization, PhD dissertation, California Inst. Technol., Pasadena, 2000.

[33] S. Prajna, A. Papachristodoulou, P. Seiler, SOSTOOLS: Sum of Squares Optimization toolbox for MATLAB, User's Guide, 2004.

[34] S.K. Nguang, Robust nonlinear H_∞ output feedback control, IEEE Transactions on Automatic Control 41 (7) (1996) 1005–1007.

[35] S.K. Nguang, W. Assawinchaichote, H_∞ filtering for fuzzy dynamical systems with D-stability constraints, IEEE Transactions on Circuits and Systems. I, Regular Papers 50 (11) (Nov. 2003) 1503–1508.

Chapter

Global stabilization of fuzzy polynomial discrete-time nonlinear systems

CHAPTER OUTLINE

7.1 INTRODUCTION

Lyapunov's method has been applied by many researchers in the past century to investigate the stability of nonlinear systems [1]. To show that a system is stable in the sense of Lyapunov, a positive definite function of the system states, which decreases along system trajectories (Lyapunov function), should be found. In recent years, advances in convex programming and numerical solutions for linear matrix inequalities (LMIs) [2] and sum of squares (SOS) [3,4] have significantly attracted researchers in developing Lyapunov functions for stabilization of nonlinear systems. These nonlinear systems are represented by various types of models such as the Takagi–Sugeno (T-S) fuzzy model, polynomial fuzzy model, etc. The Takagi–Sugeno (T-S) fuzzy model has been shown as a universal approximator of any smooth nonlinear system, and a huge amount of research results related to stability analysis and controller synthesis, based on this model, have been published (see [5] and references therein). In many of them, the stabilization problem leads to numerical solution of LMIs [6–12,33].

Another class of models that has gained considerable attention from researchers in controller synthesis and stability analysis of nonlinear systems is the polynomial fuzzy model [34–48]. Polynomial fuzzy models are a generalization of the T-S fuzzy models with smaller number of IF–THEN rules,

Analysis and Synthesis of Polynomial Discrete-Time Systems. DOI: 10.1016/B978-0-08-101901-6.00007-4

but they are more effective in representing nonlinear systems [34,36]. Furthermore, stability of polynomial fuzzy models usually leads to a convex optimization problem in terms of sum of squares, which can be solved using convex programming.

Several progresses have been made in control synthesis and stability analysis of polynomial fuzzy systems using the SOS approach, and several directions have been explored in the literature. For example, a polynomial fuzzy observer has been investigated in [42], and in [44], a parameter-dependent polynomial fuzzy model has been considered, and a polynomial controller has been designed. In [41,43], and [45], respectively, problems of output-feedback stabilization, regulation, and tracking have been investigated for polynomial fuzzy systems, and SOS-based conditions have been derived. In the new direction of using the SOS approach, in [38–40,47], the shapes of membership functions are approximated as polynomial functions and employed to facilitate the stability analysis and to reduce the conservativeness in control design.

All aforementioned studies deal with continuous-time polynomial fuzzy systems. Recently in [48], discrete-time polynomial T-S fuzzy models have been introduced and demonstrated to be more effective than discrete-time T-S fuzzy models in representing nonlinear systems. However, the method proposed in [48] cannot guarantee a Lyapunov function to be a radially unbounded polynomial function, and hence the global stability cannot be assured.

In this chapter, we propose a method that guarantees to render a Lyapunov function that is a radially unbounded polynomial function. To the best of our knowledge, there exist no results yet on guaranteeing polynomial Lyapunov functions. With a polynomial Lyapunov function, the global stability of discrete-time polynomial fuzzy systems with a state feedback controller can be guaranteed. The problem of nonconvexity is solved by incorporating an integrator into the controller structure. This solution has been proposed for the first time in [49] for discrete-time polynomial nonlinear systems, but it cannot guarantee the global stability. In this study, this is developed for polynomial fuzzy systems in a more general framework to ensure the global stability. Finally, sufficient conditions for the existence of state feedback controller are given in terms of polynomial matrix inequalities, which are solved via SOSTOOLS [4].

Notation: $(\cdot)^T$, $(\cdot)^{-T}$, and $\bullet$ denote the transpose, the transpose of the inverse, and the symmetric block in a symmetric matrix, respectively; $\mu_i(k) = \mu_i(\upsilon(k))$, $x_k = x(k)$, $x_{c,k} = x_c(k)$, and $\check{x}_k = \check{x}(x_k)$.

7.2 SYSTEM DESCRIPTION AND PROBLEM FORMULATION

Consider the nonlinear system

$$x(k+1) = f(x(k)) + g(x(k))u(k), \tag{7.1}$$

where f and g are nonlinear functions, and $x(k) \in \Re^n$ and $u(k) \in \Re^m$ are the state and control signals, respectively. By using the concept of sector non-linearity [50], system (7.1) can be converted to discrete-time polynomial fuzzy system, which is described by the following rules.

Plant Rule i**:** IF $\upsilon_1(k)$ is J_1^i AND $\cdots$ AND $\upsilon_q(k)$ is J_q^i, THEN

$$x_{k+1} = \sum_{i=1}^{r} \mu_i(\upsilon(k))\{A_i(x_k)\check{x}_k + B_i(x_k)u_k\}, \tag{7.2}$$

where i denotes the ith fuzzy rule ($i = 1, \ldots, r$), $\upsilon_1(k), \ldots, \upsilon_q(k)$ are the premise variables, q is the number of premise variables, and $J_1^i, \ldots, J_q^i$ are the premise variable components. By using a center-average defuzzifier, product inference, and singleton fuzzifier, the ith membership function is defined as $\mu_i(\upsilon(k)) = \frac{\chi_i(\upsilon(k))}{\sum_{\ell=1}^{r} \chi_\ell(\upsilon(k))} \in [0, 1]$, where $\chi_i(\upsilon(k)) = \prod_{t=1}^{q} J_t^i(\upsilon_t(k))$ and $\sum_{i=1}^{r} \mu_i(\upsilon(k)) = 1$. The matrices $A_i(x_k)$ and $B_i(x_k)$ have appropriate dimensions; $\check{x}_k \in \Re^N$ is a column vector such that $\check{x}_k = T(x_k)x_k$, where $T(x_k)$ is a polynomial transformation matrix from x_k to $\check{x}_k$. Note that $\check{x}_k$ is not unique and gives flexible options to construct the polynomial fuzzy model (7.2). The model of (7.2) can be expressed as follows:

$$x_{k+1} = A(\mu, x_k)\check{x}_k + B(\mu, x_k)u_k, \tag{7.3}$$

where $A(\mu, x_k) = \sum_{i=1}^{r} \mu_i(\upsilon(k))A_i(x_k)$ and $B(\mu, x_k) = \sum_{i=1}^{r} \mu_i(\upsilon(k)) \times B_i(x_k)$.

Using the parallel distributed compensation, the overall fuzzy polynomial controller is given as

$$u_k = \sum_{i=1}^{r} \mu_i(\upsilon(k))K_i(x_k)\check{x}_k = K(\mu, x_k)\check{x}_k. \tag{7.4}$$

The fuzzy system (7.3) with controller (7.4) is expressed as

$$x_{k+1} = A_{cl}(\mu, x_k)\check{x}_k, \tag{7.5}$$

where

$$A_{cl}(\mu, x_k) = \sum_{i=1}^{r}\sum_{j=1}^{r} \mu_i(k)\mu_j(k)\left(A_i(x_k) + B_i(x_k)K_j(x_k)\right). \quad (7.6)$$

Before stating our main result, in the sequel, we will review an existing result on the stabilization of the polynomial fuzzy system (7.2) using a state feedback controller (7.4).

Theorem 19. *[48] The control system consisting of (7.3) and (7.4) is stable if there exist a symmetric polynomial matrix $P(\tilde{x}_k)$, polynomial matrices $L_i(x_k)$, and polynomial functions $\sigma_{ijg}(x_k)$ such that (7.7)–(7.10) are satisfied, where $\epsilon_1(x_k)$ and $\epsilon_{2i}(x_k)$ are nonnegative polynomial functions such that $\epsilon_1(x_k) > 0$ for $x_k \neq 0$ and $\epsilon_{2i}(x_k) \geq 0$.*

$$y_1^T\left(P(\tilde{x}_k) - \epsilon_1(x_k)I\right)y_1 \text{ is an SOS}, \quad (7.7)$$

$$y_2^T\left(S_{ii}(x_k) - D_i(x_k)\right)y_2 \text{ is an SOS}, \quad (7.8)$$

$$y_3^T\left(S_{ij}(x_k) + S_{ij}(x_k)\right)y_3 \text{ is an SOS}, \quad (7.9)$$

$$\sigma_{ijl}(x_k) \text{ is an SOS}, \quad \forall\, i, j, l, \quad (7.10)$$

where $\tilde{x}_k$ is the substate vector such that the corresponding rows of $B_i(x_k)$) are zero rows, $S_{ij}(x_k) = \begin{bmatrix} P(\tilde{x}_k) + \sum_{l=1}^{n}\sigma_{ijl}(x_k)d_l(x_l)I & \bullet \\ T\left(\tilde{A}(x_k)x_k\right)\Omega_{ij}(x_k) & P\left(\tilde{A}(x_k)x_k\right) \end{bmatrix}$, $D_i(x_k) = \begin{bmatrix} -\epsilon_{2i}(x_k)I & \bullet \\ 0 & 0 \end{bmatrix}$, $\Omega_{ij}(x_k) = A_i(x_k)P(\tilde{x}_k) + B_i(x_k)L_j(x_k)$, $d_l(x_l) = \left(x_l(k) - x_{l,min}\right)\left(x_l(k) - x_{l,max}\right)$, and $y_1 \in \Re^N$ and $y_2, y_3 \in \Re^{2N}$ are vectors independent of x_k. In addition, if (7.7)–(7.10) hold with $\epsilon_{2i}(x_k) > 0$ for $x \neq 0$, then the equilibrium $x = 0$ is asymptotically stable.

Stabilizing feedback gains $K_i(x_k)$ can be obtained from $P(\tilde{x}_k)$ and $L_i(x_k)$ as

$$K_i(x_k) = L_i(x_k)P^{-1}(\tilde{x}_k). \quad (7.11)$$

Remark 35. In [48], the Lyapunov function is given as follows:

$$V(x_k) = \check{x}_k^T P^{-1}(\tilde{x}_k)\check{x}_k, \quad (7.12)$$

where $P(\tilde{x}_k)$ is a polynomial positive definite matrix that satisfies the SOS constraints given in Theorem 19. Clearly, because of the inversion of $P(\tilde{x}_k)$, the Lyapunov function given in (7.12) may not be a radially unbounded polynomial function, and hence the global stability cannot be guaranteed. Also in [48], $P(\tilde{x}_k)$ has been restricted to be a function of $\tilde{x}_k$, where $\tilde{x}_k$

is the substate vector such that the corresponding rows of $B_i(x_k))$ are zero rows. In this paper, we relax this limitation to allow $P(\cdot)$ to be a function of all the states, which makes it less conservative.

In general, when $P(x_k)$ is a positive definite polynomial matrix, $P^{-1}(x_k)$ is not always polynomial. In this study, we look for a particular form of positive definite polynomial matrix $P(x_k)$ whose inverse is also a positive definite polynomial matrix. In doing so, it will yield a Lyapunov function $V(x_k)$ to be radially unbounded, and the global stability is guaranteed. We express $P(x_k) = \hat{G}^T(x_k)\hat{P}^{-1}\hat{G}(x_k)$, where $\hat{P}$ is a positive definite matrix, and $\hat{G}(x_k)$ is selected as a lower triangular polynomial matrix

$$\hat{G}(x_k) = \begin{bmatrix} \hat{G}_{11} & 0 & \cdots & 0 \\ \hat{G}_{21}(x_k) & \hat{G}_{22} & 0 & 0 \\ \vdots & & \ddots & 0 \\ \hat{G}_{N1}(x_k) & \hat{G}_{N2}(x_k) & \cdots & \hat{G}_{NN} \end{bmatrix}, \tag{7.13}$$

where $\hat{G}_{21}(x_k), \ldots, \hat{G}_{N2}(x_k)$ are polynomial functions, and $\hat{G}_{11}, \ldots, \hat{G}_{NN}$ are scalars. Since $\hat{G}(x_k)$ is lower triangular matrix, $\det\left(\hat{G}(x_k)\right)$ is a scalar, and $\hat{G}^{-1}(x_k) = \frac{1}{det\left(\hat{G}(x_k)\right)} adj(\hat{G}(x_k))$, which is a polynomial matrix. So, $P^{-1}(x_k) = \hat{G}^{-T}(x_k)\hat{P}\hat{G}^{-1}(x_k)$ is a polynomial matrix. Consequently, the Lyapunov function $V(x_k) = x_k^T\hat{G}^{-T}(x_k)\hat{P}\hat{G}^{-1}(x_k)x_k$ is always a polynomial function of the system states and radially unbounded. This Lyapunov function ensures the global stability of the closed-loop system.

Remark 36. Based on the Cholesky decomposition, for any given positive definite matrix P, we can always decompose P into $\hat{G}^{-T}\hat{P}\hat{G}^{-1}$, where $\hat{P}$ is another positive definite matrix, and $\hat{G}^{-1}$ is a lower (upper) triangular matrix. In linear systems, expressing P as $\hat{G}^{-T}\hat{P}\hat{G}^{-1}$ does not affect the stability results.

Before proceeding to the controller design, we recall the following inequality lemma, which will be used in the sequel.

Lemma 4. *[51] For a given matrix $\hat{G}$, if $\hat{P} > 0$, then $\hat{G}^T + \hat{G} - \hat{P} \leq \hat{G}^T\hat{P}^{-1}\hat{G}$.*

7.3 MAIN RESULTS

In this section, we propose a theorem for global stabilization of discrete-time polynomial fuzzy systems (7.3) with state feedback control. This theorem offers sufficient convex conditions in the form of polynomial matrix inequalities. The problem of nonconvexity is resolved using the integrator approach. In this method, an integrator is incorporated into the controller structure as follows:

$$\begin{aligned} x_{c,k+1} &= x_{c,k} + A_c(\mu_k, x_k, x_{c,k}), \\ u_k &= x_{c,k}, \end{aligned} \tag{7.14}$$

where $x_{c,k}$ is an additional state or controller state, and $A_c(\mu, x_k, x_{c,k}) = \sum_{i=1}^{r} \mu_i(\upsilon(k)) A_{c_i}(x_k, x_{c,k})$ is the input of the integrator. To design the controller, we should find $A_{c_i}(x_k, x_{c,k})$ such that the system is globally stable. From (7.3) and (7.14) the dynamics of the closed-loop system is given by

$$\bar{x}_{k+1} = \hat{A}_{cl}(\mu, \bar{x}_k)\hat{x}_k, \tag{7.15}$$

where $\hat{x}_k^T = \begin{bmatrix} \check{x}_k^T & x_{c,k}^T \end{bmatrix}$, $\bar{x}_k^T = \begin{bmatrix} x_k^T & x_{c,k}^T \end{bmatrix}$, $\hat{x}_k = \hat{T}(x_k)\bar{x}_k$, $\hat{T}(x_k) = diag\{T(x_k), 1\}$, $\hat{A}_{cl}(\mu, \bar{x}_k) = \sum_{i=1}^{r}\sum_{j=1}^{r} \mu_i(k)\mu_j(k)\hat{A}_i(x_k) + \hat{B}\hat{A}_{c_j}(\bar{x}_k)$, $\hat{A}_{c_i}(\bar{x}_k)\hat{x}_k = A_{c_i}(x_k, x_{c,k})$, $\hat{A}_i(x_k) = \begin{bmatrix} A_i(x_k) & B_i(x_k) \\ 0_{1\times n} & 1 \end{bmatrix}$, and $\hat{B} = \begin{bmatrix} 0_{n\times 1} \\ 1 \end{bmatrix}$.

Theorem 20. *The polynomial fuzzy system described by (7.3) with controller (7.14) is globally asymptotically stable if there exist a positive definite matrix $\hat{P}$, a lower triangular polynomial matrix function $\hat{G}(x_k)$, and polynomial functions $\epsilon_{1i}(\bar{x}_k) > 0$ and $\epsilon_{1i}(\bar{x}_k) > 0$ such that the following conditions hold:*

$$\begin{cases} y^T\left(\hat{\Lambda}_{ii} - \epsilon_{i1}(\bar{x}_k)I\right)y & \text{is an SOS}, \\ & 1 \le i \le r, \\ y^T\left(\frac{1}{r-1}\hat{\Lambda}_{ii} + \frac{1}{2}\left(\hat{\Lambda}_{ij} + \hat{\Lambda}_{ji}\right) - \epsilon_{2ij}(\bar{x}_k)I\right)y & \text{is an SOS}, \\ & 1 \le i \ne j \le r, \end{cases} \tag{7.16}$$

where

$$\hat{\Lambda}_{ij} = \begin{bmatrix} \hat{P} & \bullet \\ \hat{T}(x_{k+1})\hat{\varphi}_{ij}(\bar{x}_k) & \hat{G}(x_{k+1}) + \hat{G}^T(x_{k+1}) - \hat{P} \end{bmatrix}, \tag{7.17}$$

$$\hat{\varphi}_{ij}(\bar{x}_k) = \hat{A}_i(\bar{x}_k)\hat{G}(x_k)) + \hat{B}_i\hat{L}_j(\bar{x}_k), \tag{7.18}$$

and $y \in \Re^{2N}$ is a vector independent of x_k. Further, the controller parameters can be obtained from $\hat{G}(x_k)$ and $\hat{L}_i(\hat{x}_k)$ as

$$\hat{A}_{c_i}(\bar{x}_k) = \hat{L}_i(\bar{x}_k)\hat{G}^{-1}(x_k) \tag{7.19}$$

for $i = 1, \ldots, r$.

Proof. Consider the following positive definite polynomial Lyapunov function:

$$V(\bar{x}_k) = \hat{x}_k^T \hat{G}^{-T}(x_k)\hat{P}\hat{G}^{-1}(x_k)\hat{x}_k. \tag{7.20}$$

Taking the forward difference of $V(\bar{x}_k)$ gives

$$\begin{aligned} \Delta V(\bar{x}_k) &= V(\bar{x}_{k+1}) - V(\bar{x}_k) \\ &= \hat{x}_{k+1}^T \hat{G}^{-T}(x_{k+1})\hat{P}\hat{G}^{-1}(x_{k+1})\hat{x}_{k+1} \\ &- \hat{x}_k^T \hat{G}^{-T}(x_k)\hat{P}\hat{G}^{-1}(x_k)\hat{x}_k. \end{aligned} \tag{7.21}$$

Substituting the value of $\hat{x}_{k+1}$ from (7.15) into (7.21) gives

$$\begin{aligned} &\Delta V(\bar{x}_k) \\ &= \bar{x}_{k+1}^T \hat{T}^T(x_{k+1})\hat{G}^{-T}(x_{k+1})\hat{P}\hat{G}^{-1}(x_{k+1})\hat{T}(x_{k+1})\bar{x}_{k+1} \\ &- \hat{x}_k^T \hat{G}^{-T}(x_k)\hat{P}\hat{G}^{-1}(x_k)\hat{x}_k \\ &= \hat{x}_k^T \Big[\hat{A}_{cl}^T(\mu, \bar{x}_k)\hat{T}^T(x_{k+1})\hat{G}^{-T}(x_{k+1})\hat{P}\hat{G}^{-1}(x_{k+1})\hat{T}(x_{k+1})\hat{A}_{cl}(\mu, \bar{x}_k) \\ &- \hat{G}^{-T}(x_k)\hat{P}\hat{G}^{-1}(x_k)\Big]\hat{x}_k \end{aligned}$$

The closed-loop fuzzy system is globally asymptotically stable if $\Delta V(\bar{x}_k) < 0$ or, equivalently, if the following inequality holds:

$$\begin{aligned} &\Big[\hat{A}_{cl}^T(\mu, \bar{x}_k)\hat{T}^T(x_{k+1})\hat{G}^{-T}(x_{k+1})\hat{P}\hat{G}^{-1}(x_{k+1})\hat{T}(x_{k+1})\hat{A}_{cl}(\mu, \bar{x}_k) \\ &- \hat{G}^{-T}(x_k)\hat{P}\hat{G}^{-1}(x_k)\Big] < 0. \end{aligned} \tag{7.22}$$

Multiplying right- and left-hand sides of (7.22) by $\hat{G}^T(x_k)$ and $\hat{G}(x_k)$, respectively, we have

$$\hat{\varphi}^T(\mu, \bar{x}_k)\hat{T}^T(x_{k+1})\hat{G}^{-T}(x_{k+1})\hat{P}\hat{G}^{-1}(x_{k+1})\hat{T}(x_{k+1})\hat{\varphi}(\mu, \bar{x}_k) - \hat{P} < 0, \tag{7.23}$$

where $\hat{\varphi}(\mu, \bar{x}_k) = \sum_{i=1}^{r}\sum_{j=1}^{r}\mu_i(k)\mu_j(k)\hat{A}_i(x_k) + \hat{B}\hat{L}_j(\bar{x}_k)$. Applying the Schur complement to inequality (7.23) leads to

$$\begin{bmatrix} \hat{P} & \bullet \\ \hat{T}(x_{k+1})\hat{\varphi}(\mu, \bar{x}_k) & \hat{G}^{-T}(x_{k+1})\hat{P}\hat{G}^{-1}(x_{k+1}) \end{bmatrix} > 0. \tag{7.24}$$

Using *Lemma* 4, inequality (7.24) holds if

$$\Lambda = \begin{bmatrix} \hat{P} & \bullet \\ \hat{T}(x_{k+1})\hat{\varphi}(\mu, \bar{x}_k) & \hat{G}(x_{k+1}) + \hat{G}^T(x_{k+1}) - \hat{P} \end{bmatrix} > 0, \quad (7.25)$$

where $\Lambda = \sum_{i=1}^{r}\sum_{j=1}^{r}\mu_i(\upsilon(k))\mu_j(\upsilon(k))\Lambda_{ij}$. Following [8], inequality (7.25) can be expressed as $\hat{\Lambda}_{ii} > 0$ for $1 \le i \le r$ and $\frac{1}{r-1}\hat{\Lambda}_{ii} + \frac{1}{2}\left(\hat{\Lambda}_{ij} + \hat{\Lambda}_{ji}\right) > 0$ for $1 \le i \ne j \le r$, which can be satisfied by conditions in (7.16). This completes the proof. □

Remark 37. Theorem 20 provides sufficient convex conditions for the existence of a state feedback control. The existence of $\hat{G}(x_{k+1})$ or, equivalently, $\hat{G}\left(\sum_{i=1}^{r}\mu_i(\upsilon(k))\{A_i(x_k)\check{x}_k + B_i(x_k)x_{c.k}\}\right)$ does not make conditions in (7.16) nonconvex because $x_{c.k}$ is the controller state. Further, $\hat{G}(x_k)$ can be assumed either a lower or upper triangular polynomial matrix, depending on which gives a feasible solution.

Remark 38. A column vector $\check{x}_k = \hat{T}(x_k)x_k$ whose entries are the monomials (in x_k) has been chosen in a more general form compared with results in [48]. This plays an important role in construction of the SOS design. A greater number of monomials in $\check{x}_k$ leads to less conservative results. However, the computational burden increases.

7.4 SIMULATION EXAMPLES

Example 1. Let us consider the following discrete time nonlinear system:

$$\begin{aligned} x_1(k+1) &= -x_1(k)\sin(x_1(k) + 0.1x_2^2(k) + u(k), \\ x_2(k+1) &= 0.2x_1(k) + x_2(k). \end{aligned} \quad (7.26)$$

Using the technique described in [50] with the premise variable $\upsilon(k) = x_1(k)$, the nonlinear model (7.26) can be expressed by the polynomial fuzzy model structure given by

$$\begin{aligned} &\begin{bmatrix} x_1(k+1) \\ x_2(k+1) \end{bmatrix} \\ &= \mu_1(\upsilon(k))\left\{\begin{bmatrix} 1 & 0.1x_2(k) \\ 0.2 & 1 \end{bmatrix}\begin{bmatrix} x_1(k) \\ x_2(k) \end{bmatrix} + \begin{bmatrix} 1 \\ 0 \end{bmatrix}u(k)\right\} \\ &+ \mu_2(\upsilon(k))\left\{\begin{bmatrix} -1 & 0.1x_2(k) \\ 0.2 & 1 \end{bmatrix}\begin{bmatrix} x_1(k) \\ x_2(k) \end{bmatrix} + \begin{bmatrix} 1 \\ 0 \end{bmatrix}u(k)\right\}, \end{aligned} \quad (7.27)$$

where $\mu_1(\upsilon(k)) = \frac{1}{2}(1 + \sin(\upsilon(k)), \mu_2(\upsilon(k)) = 1 - \mu_1(\upsilon(k))$, and $-\frac{\pi}{2} \le \upsilon(k) \le \frac{\pi}{2}$.

Define

$$T(x_2(k)) = \begin{bmatrix} 1 & 0 \\ 0 & 1 \\ 0 & x_2(k) \end{bmatrix} \text{ and } \check{x}_k = T(x_2(k)). \begin{bmatrix} x_1(k) \\ x_2(k) \end{bmatrix} = \begin{bmatrix} x_1(k) \\ x_2(k) \\ x_2^2(k) \end{bmatrix}. \tag{7.28}$$

Applying Theorem 19, a feasible solution is achieved by selecting $\epsilon_1(x_k) = 0.01$ and the degree of each term in $P(x_2(k))$ and $L_i(x_k)$ equals 6 and 8, respectively. Here, $P(x_2(k))$ is a symmetrical 3×3 polynomial matrix whose (i, j)th entry is given by

$$p_{ij}(x_2(k)) = p_{0.ij} + p_{1.ij}x_2 + p_{2.ij}x_2^2 + p_{3.ij}x_2^3 + p_{4.ij}x_2^4 + p_{5.ij}x_2^5 + p_{6.ij}x_2^6, \tag{7.29}$$

where $i, j = 1, 2, 3$. Further, the jth element of $L_i(x_k)$, denoted as $L_{ij}(x_k)$ (for $i = 1, 2$ and $j = 1, 2, 3$), is given by

$$L_{ij}(x_k) = l_{0.ij} + l_{11.ij}x_1 + l_{12.ij}x_2 + l_{21.ij}x_1^2 + l_{22.ij}x_2^2 + l_{23.ij}x_1x_2 + \cdots + l_{81.ij}x_1^8 + l_{82.ij}x_2^8. \tag{7.30}$$

The values of the polynomial elements in matrices $P(x_2(k))$ and $L_i(x_k)$ are omitted here due to limitation of space.

It is known that the system is globally stable when the Lyapunov function is radially unbounded, i.e., $V \to \infty$ as $\|x\| \to \infty$. In this example, the selected Lyapunov function $\check{x}_k^T P^{-1}(x_2)\check{x}_k$ is a rational function with respect to x_2. Fig. 7.1 shows its contour curves ($V(x) = \alpha$). These contours are not closed curves, so the radially unboundedness condition is not satisfied [52]. When the contours are not closed curves, it is possible for the state trajectories ($x_2(k)$ in this case) to drift away from the equilibrium point in some situations as shown in Fig. 7.1. Now, we design the controller via Theorem 20. By appending an integrator into (7.27) and selecting $T(x_k) = I$, system (7.27) is represented in the form of (7.15) with $\bar{x}_k^T = \begin{bmatrix} x_1(k) & x_2(k) & x_c(k) \end{bmatrix}$,

$$\hat{A}_1(x_k) = \begin{bmatrix} -1 & 0.1x_2(k) & 1 \\ 0.2 & 1 & 0 \\ 0 & 0 & 1 \end{bmatrix}, \quad \hat{B}_1 = \begin{bmatrix} 0 \\ 0 \\ 1 \end{bmatrix},$$

$$\hat{A}_2(x_k) = \begin{bmatrix} 1 & 0.1x_2(k) & 1 \\ 0.2 & 1 & 0 \\ 0 & 0 & 1 \end{bmatrix}, \quad \hat{B}_1 = \begin{bmatrix} 0 \\ 0 \\ 1 \end{bmatrix}.$$

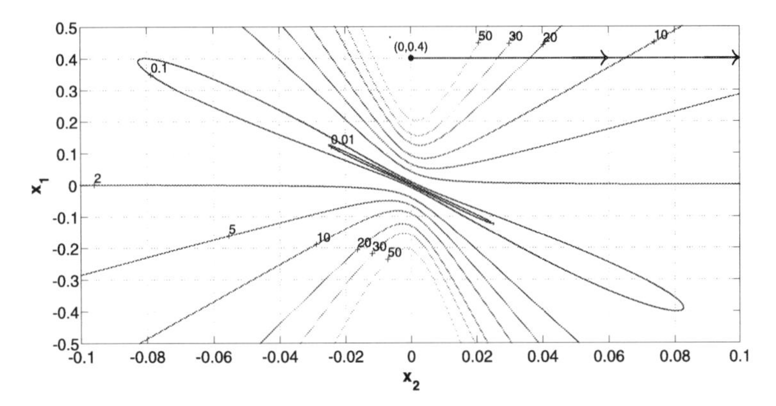

FIGURE 7.1 Contours of the Lyapunov function $V(x_1, x_2)$ using Theorem 19.

The parameters of ϵ_1 and ϵ_2 are selected to be 0.01. The structure of $\hat{G}(x_k)$, $\hat{L}_1(\hat{x}_k)$, and $\hat{L}_2(\hat{x}_k)$ is given by

$$\hat{G}(x_k) = \begin{bmatrix} \hat{G}_{11} & 0 & 0 \\ \hat{G}_{21}(x_k) & \hat{G}_{22} & 0 \\ \hat{G}_{31}(x_k) & \hat{G}_{32}(x_k) & \hat{G}_{33} \end{bmatrix},$$

$$\hat{L}_1(\bar{x}_k) = \begin{bmatrix} \hat{L}_{11}(\bar{x}_k) & \hat{L}_{12}(\bar{x}_k) & \hat{L}_{13}(\bar{x}_k) \end{bmatrix},$$

$$\hat{L}_2(\bar{x}_k) = \begin{bmatrix} \hat{L}_{21}(\bar{x}_k) & \hat{L}_{22}(\bar{x}_k) & \hat{L}_{23}(\bar{x}_k) \end{bmatrix}.$$

The degrees of each polynomial element in polynomial matrices $\hat{G}(x_k)$ and $\hat{L}_i(\bar{x}_k)$ are selected to be 7 and 8, respectively. With these parameters, a feasible solution is obtained. We omit the values of the polynomial elements in $\hat{G}(x_k)$, $\hat{L}_1(\bar{x}_k)$, and $\hat{L}_2(\bar{x}_k)$ due to limitation of space. Fig. 7.2 shows the contour curves for the proposed Lyapunov function in three subplots: (a) $V(x_1, 0, z) = \alpha$, (b) $V(0, x_2, z) = \alpha$, and (c) $V(x_1, x_2, 0) = \alpha$ for $\alpha = 0.1, 1, 5, 10, 20, 30$. Note that, in all of the plots shown in Fig. 7.2, contour curves are closed. Thus, if the states tend to infinity in any direction, V increases and tends to infinity. Hence, the Lyapunov function is radially unbounded, and the closed-loop system is globally stable.

Fig. 7.3 shows the state responses of the closed-loop system considering the proposed controller with initial values $x(0) = (0.4, 0, 0)$. We can observe that eventually the state responses converge to the origin. For comparison, the same initial values $x(0) = (0.4, 0)$ are investigated for the closed-loop system considering the controller designed in [48]. Unlike the proposed controller, the states diverge away from the initial values. This confirms that the closed-loop system with controller designed as [48] is not globally stable.

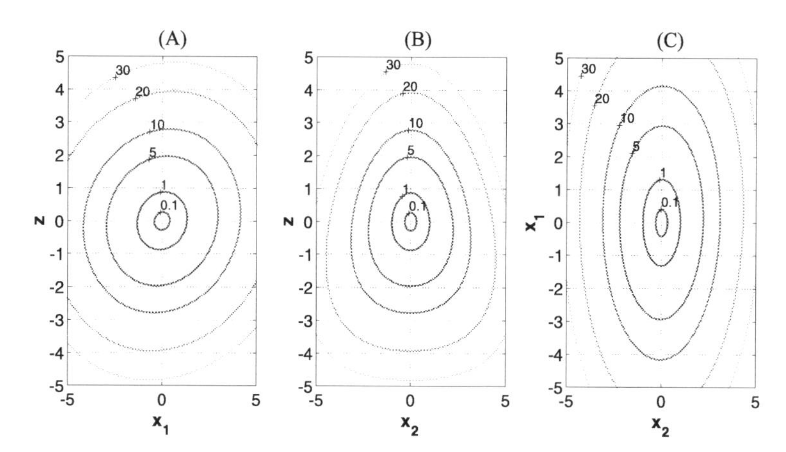

■ **FIGURE 7.2** Contours of the Lyapunov function. (A): $V(x_1, 0, z)$, (B): $V(0, x_2, z)$, and (C): $V(x_1, x_2, 0)$.

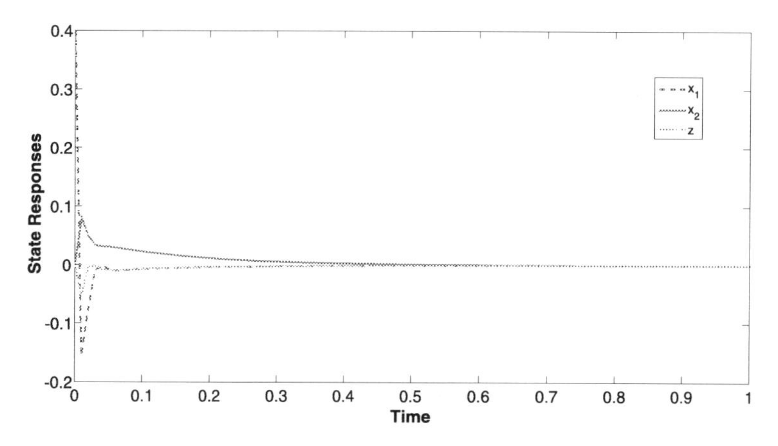

■ **FIGURE 7.3** Closed-loop system states with the proposed controller.

Example 2. Consider the following discrete-time nonlinear pendulum system:

$$\begin{aligned} x_1(k+1) &= x_1(k) + 0.1x_2(k) \\ x_2(k+1) &= 0.1x_1(k)\sin(x_1(k) + 0.1x_2^2(k)x_1(k) + x_2(k) + 0.1u(k) \end{aligned} \tag{7.31}$$

Employing [38] with the premise variable $\upsilon(k) = x_1(k)$, (7.31) can be expressed in the form of (7.3) with

$$A_1(x(k)) = \begin{bmatrix} 1 & 0.1 \\ 0.1 & 0.1x_1(k)x_2(k) \end{bmatrix},\ B_1(x(k)) = \begin{bmatrix} 0 \\ 0.1 \end{bmatrix},$$

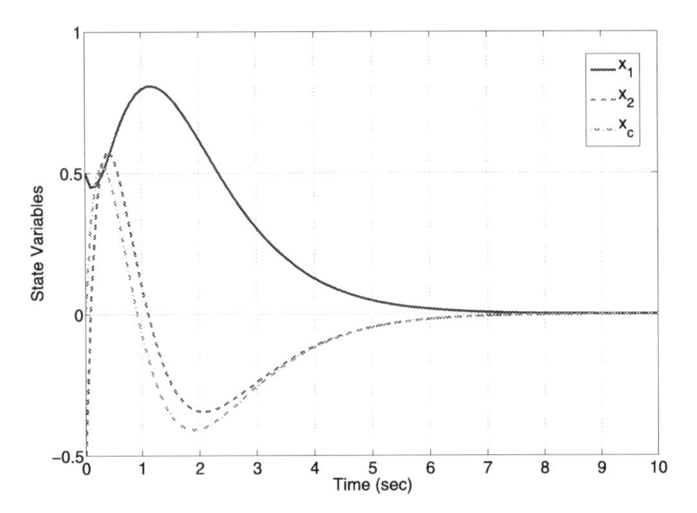

FIGURE 7.4 State responses.

$$A_2(x(k)) = \begin{bmatrix} 1 & 0.1 \\ -0.1 & 0.1x_1(k)x_2(k) \end{bmatrix}, B_2(x(k)) = \begin{bmatrix} 0 \\ 0.1 \end{bmatrix},$$

$\mu_1(\upsilon(k)) = \frac{1}{2}(1+\sin(\upsilon(k)), \mu_2(\upsilon(k)) = 1-\mu_1(\upsilon(k))$, and $-\frac{\pi}{2} \leq \upsilon(k) \leq \frac{\pi}{2}$.

We append an integrator and apply Theorem 19 with $T(x_k) = I, \epsilon_1 = 0.01$, and $\epsilon_2 = 0.01$. The structure of $\hat{G}(x_k)$, $\hat{L}_1(\hat{x}_k)$, and $\hat{L}_2(\hat{x}_k)$ is given by

$$\hat{G}(x_k) = \begin{bmatrix} \hat{G}_{11} & 0 & 0 \\ \hat{G}_{21}(x_k) & \hat{G}_{22} & 0 \\ \hat{G}_{31}(x_k) & \hat{G}_{32}(x_k) & \hat{G}_{33} \end{bmatrix},$$

$$\hat{L}_1(\bar{x}_k) = \begin{bmatrix} \hat{L}_{11}(\bar{x}_k) & \hat{L}_{12}(\bar{x}_k) & \hat{L}_{13}(\bar{x}_k) \end{bmatrix},$$

$$\hat{L}_2(\bar{x}_k) = \begin{bmatrix} \hat{L}_{21}(\bar{x}_k) & \hat{L}_{22}(\bar{x}_k) & \hat{L}_{23}(\bar{x}_k) \end{bmatrix},$$

where $\bar{x}^T(k) = \begin{bmatrix} x_1^T(k) & x_2^T(k) & x_c^T(k) \end{bmatrix}$. The degrees of each polynomial element in polynomial matrices $\hat{G}(x_k)$ and $\hat{L}_i(\bar{x}_k)$ are selected to be 8 and 10, respectively. With these parameters, a feasible solution is obtained. The values of the polynomial elements in $\hat{G}(x_k)$, $\hat{L}_1(\bar{x}_k)$, and $\hat{L}_2(\bar{x}_k)$ are too large to be included here. Fig. 7.4 shows the state responses of the closed-loop system considering the proposed controller with initial values $x(0) = (0.5, -0.5)$ and $x_c(0) = 0$. We can observe that eventually the state responses converge to the origin.

7.5 CONCLUSION

In this chapter, we derived sufficient conditions for the existence of a state feedback controller for discrete-time polynomial fuzzy systems in terms of polynomial matrix inequalities. Using a polynomial Lyapunov function, the global stability of nonlinear systems, represented by polynomial fuzzy models, are established. It has been demonstrated that by incorporating the integrator into the controller design, a convex solution can be obtained. Therefore, the problem can be solved using semidefinite programming. The effectiveness of the proposed design techniques has been demonstrated considering a numerical example, and it was shown that the proposed controller design method can achieve global stability.

REFERENCES

[1] H.K. Khalil, Nonlinear Systems, 3rd edition, Prentice Hall, 2002.

[2] S. Boyd, L. El Ghaoui, E. Feron, V. Balakrishnan, Linear Matrix Inequalities in System and Control Theory, Society for Industrial and Applied Mathematics (SIAM), 1994.

[3] A. Papachristodoulou, S. Prajna, A tutorial on sum of squares techniques for systems analysis, in: Proceedings of the American Control Conference, 2005, vol. 4, June 2005, pp. 2686–2700.

[4] S. Prajna, A. Papachristodoulou, P. Parrilo, Introducing SOSTOOLS: a general purpose sum of squares programming solver, in: Proceedings of the 41st IEEE Conference on Decision and Control, 2002, vol. 1, Dec. 2002, pp. 741–746.

[5] G. Feng, A survey on analysis and design of model-based fuzzy control systems, IEEE Transactions on Fuzzy Systems 14 (5) (Oct. 2006) 676–697.

[6] S.K. Nguang, P. Shi, Fuzzy H_∞ output feedback control of nonlinear systems under sampled measurements, Automatica 39 (12) (2003) 2169–2174.

[7] S.K. Nguang, P. Shi, S. Ding, Fault detection for uncertain fuzzy systems: an LMI approach, IEEE Transactions on Fuzzy Systems 15 (6) (Dec. 2007) 1251–1262.

[8] H. Tuan, P. Apkarian, T. Narikiyo, Y. Yamamoto, Parameterized linear matrix inequality techniques in fuzzy control system design, IEEE Transactions on Fuzzy Systems 9 (2) (2001) 324–332.

[9] G. Feng, Stability analysis of discrete-time fuzzy dynamic systems based on piecewise Lyapunov functions, IEEE Transactions on Fuzzy Systems 12 (1) (Feb. 2004) 22–28.

[10] Y. Zheng, H. Fang, H. Wang, Takagi–Sugeno fuzzy-model-based fault detection for networked control systems with Markov delays, IEEE Transactions on Systems, Man, and Cybernetics, Part B: Cybernetics 36 (4) (Aug. 2006) 924–929.

[11] W.-J. Wang, Y.-J. Chen, C.-H. Sun, Relaxed stabilization criteria for discrete-time T-S fuzzy control systems based on a switching fuzzy model and piecewise Lyapunov function, IEEE Transactions on Systems, Man, and Cybernetics, Part B: Cybernetics 37 (3) (June 2007) 551–559.

[12] A. Kruszewski, R. Wang, T. Guerra, Nonquadratic stabilization conditions for a class of uncertain nonlinear discrete time T-S fuzzy models: a new approach, IEEE Transactions on Automatic Control 53 (2) (March 2008) 606–611.

[13] F. Rasool, D. Huang, S.K. Nguang, Robust H_∞ output feedback control of networked control systems with multiple quantizers, Journal of the Franklin Institute 349 (3) (2012) 1153–1173.

[14] S.K. Nguang, W. Assawinchaichote, P. Shi, Y. Shi, Robust H_∞ control design for uncertain fuzzy systems with Markovian jumps: an LMI approach, in: American Control Conference, 2005, pp. 1805–1810.

[15] S.K. Nguang, P. Zhang, S.X. Ding, Parity relation based fault estimation for nonlinear systems: an LMI approach, International Journal of Automation and Computing 4 (2) (2007) 164–168.

[16] S.K. Nguang, Comments on "Robust stabilization of uncertain input-delay systems by sliding mode control with delay compensation", Automatica 37 (10) (2001) 1677.

[17] S.K. Nguang, P. Shi, H_∞ output feedback control of fuzzy system models under sampled measurements, Computers and Mathematics With Applications 46 (5) (2003) 705–717.

[18] W. Assawinchaichote, S.K. Nguang, P. Shi, Fuzzy Control and Filter Design for Uncertain Fuzzy Systems, Springer, 2006.

[19] S.K. Nguang, P. Shi, On designing filters for uncertain sampled-data nonlinear systems, Systems & Control Letters 41 (5) (2000) 305–316.

[20] D. Huang, S.K. Nguang, Robust Control for Uncertain Networked Control Systems With Random Delays, Springer Science & Business Media, 2009.

[21] J. Zhang, A.K. Swain, S.K. Nguang, Robust sensor fault estimation scheme for satellite attitude control systems, Journal of the Franklin Institute 350 (9) (2013) 2581–2604.

[22] J. Zhang, P. Shi, J. Qiu, S.K. Nguang, A novel observer-based output feedback controller design for discrete-time fuzzy systems, IEEE Transactions on Fuzzy Systems 23 (1) (2015) 223–229.

[23] S.K. Nguang, P. Shi, Delay-dependent H_∞ filtering for uncertain time delay nonlinear systems: an LMI approach, IET Control Theory & Applications 1 (1) (2007) 133–140.

[24] Z. Hou, J. Luo, P. Shi, S.K. Nguang, Stochastic stability of Ito differential equations with semi-Markovian jump parameters, IEEE Transactions on Automatic Control 51 (8) (2006) 1383–1387.

[25] W. Assawinchaichote, S.K. Nguang, P. Shi, E.K. Boukas, H_∞ fuzzy state-feedback control design for nonlinear systems with stability constraints: an LMI approach, Mathematics and Computers in Simulation 78 (4) (2008) 514–531.

[26] S. Chae, S.K. Nguang, SOS based robust H_∞ fuzzy dynamic output feedback control of nonlinear networked control systems, IEEE Transactions on Cybernetics 44 (7) (2014) 1204–1213.

[27] F. Rasool, D. Huang, S.K. Nguang, Robust H_∞ output feedback control of discrete-time networked systems with limited information, Systems & Control Letters 60 (10) (2011) 845–853.

[28] S.K. Nguang, P. Shi, Stabilisation of a class of nonlinear time-delay systems using fuzzy models, in: Proceedings of the 39th IEEE Conference on Decision and Control, 2000, pp. 5–11.

[29] S. Saat, S.K. Nguang, Nonlinear H_∞ output feedback control with integrator for polynomial discrete-time systems, International Journal of Robust and Nonlinear Control 25 (2015) 1051–1065.

[30] Y. Zhang, P. Shi, S.K. Nguang, H.R. Karimi, Observer-based finite-time fuzzy H_∞ control for discrete-time systems with stochastic jumps and time-delays, Signal Processing 97 (2014) 252–261.

[31] S. Chae, F. Rasool, S.K. Nguang, A. Swain, Robust mode delay-dependent H_∞ control of discrete-time systems with random communication delays, IET Control Theory & Applications 4 (6) (2010) 936–944.

[32] Y. Zhang, P. Shi, S.K. Nguang, Y. Song, Robust finite-time H_∞ control for uncertain discrete-time singular systems with Markovian jumps, IET Control Theory & Applications 8 (12) (2014) 1105–1111.

[33] Y. Chen, M. Tanaka, K. Tanaka, H. Wang, Stability analysis and region-of-attraction estimation using piecewise polynomial Lyapunov functions: polynomial fuzzy model approach, IEEE Transactions on Fuzzy Systems 23 (4) (2015) 1314–1322.

[34] K. Tanaka, H. Yoshida, H. Ohtake, H. Wang, A sum of squares approach to stability analysis of polynomial fuzzy systems, in: American Control Conference, July 2007, pp. 4071–4076.

[35] K. Tanaka, H. Ohtake, H. Wang, Guaranteed cost control of polynomial fuzzy systems via a sum of squares approach, IEEE Transactions on Systems, Man, and Cybernetics, Part B: Cybernetics 39 (2) (April 2009) 561–567.

[36] K. Tanaka, H. Yoshida, H. Ohtake, H. Wang, A sum-of-squares approach to modeling and control of nonlinear dynamical systems with polynomial fuzzy systems, IEEE Transactions on Fuzzy Systems 17 (4) (Aug. 2009) 911–922.

[37] A. Sala, C. Ario, Polynomial fuzzy models for nonlinear control: a Taylor series approach, IEEE Transactions on Fuzzy Systems 17 (6) (Dec. 2009) 1284–1295.

[38] M. Narimani, H. Lam, SOS-based stability analysis of polynomial fuzzy-model-based control systems via polynomial membership functions, IEEE Transactions on Fuzzy Systems 18 (5) (Oct. 2010) 862–871.

[39] H. Lam, Polynomial fuzzy-model-based control systems: stability analysis via piecewise-linear membership functions, IEEE Transactions on Fuzzy Systems 19 (3) (June 2011) 588–593.

[40] H. Lam, L. Seneviratne, Stability analysis of polynomial fuzzy-model-based control systems under perfect/imperfect premise matching, IET Control Theory & Applications 5 (15) (October 2011) 1689–1697.

[41] H. Lam, Stabilization of nonlinear systems using sampled-data output-feedback fuzzy controller based on polynomial-fuzzy-model-based control approach, IEEE Transactions on Systems, Man, and Cybernetics, Part B: Cybernetics 42 (1) (Feb. 2012) 258–267.

[42] K. Tanaka, H. Ohtake, T. Seo, M. Tanaka, H. Wang, Polynomial fuzzy observer designs: a sum-of-squares approach, IEEE Transactions on Systems, Man, and Cybernetics, Part B: Cybernetics 42 (5) (Oct. 2012) 1330–1342.

[43] H. Lam, J. Lo, Output regulation of polynomial-fuzzy-model-based control systems, IEEE Transactions on Fuzzy Systems 21 (2) (April 2013) 262–274.

[44] H. Lam, L. Seneviratne, X. Ban, Fuzzy control of non-linear systems using parameter-dependent polynomial fuzzy model, IET Control Theory & Applications 6 (11) (July 2012) 1645–1653.

[45] H. Lam, H. Li, Output-feedback tracking control for polynomial fuzzy-model-based control systems, IEEE Transactions on Industrial Electronics 60 (12) (Dec. 2013) 5830–5840.
[46] H. Lam, M. Narimani, H. Li, H. Liu, Stability analysis of polynomial-fuzzy-model-based control systems using switching polynomial Lyapunov function, IEEE Transactions on Fuzzy Systems 21 (5) (Oct. 2013) 800–813.
[47] S. Chae, S.K. Nguang, SOS based robust H_∞ fuzzy dynamic output feedback control of nonlinear networked control systems, IEEE Transactions on Cybernetics 44 (7) (July 2014) 1204–1213.
[48] Y.-J. Chen, M. Tanaka, K. Tanaka, H. Ohtake, H. Wang, Discrete polynomial fuzzy systems control, IET Control Theory & Applications 8 (4) (March 2014) 288–296.
[49] S. Saat, D. Huang, S.K. Nguang, A. Hamidon, Nonlinear state feedback control for a class of polynomial nonlinear discrete-time systems with norm-bounded uncertainties: an integrator approach, Journal of the Franklin Institute 350 (7) (Sep. 2013) 1739–1752.
[50] K. Tanaka, H.O. Wang, Fuzzy Control Systems Design and Analysis: A Linear Matrix Inequality Approach, Wiley, 2001.
[51] M.C.D. Oliveora, J. Berussou, J.C. Geromel, A new discrete-time robust stability condition, Systems & Control Letters 37 (1999) 261–265.
[52] J. Slotine, W. Li, Applied Nonlinear Control, Prentice Hall, 1991.

Chapter 8

Global H_∞ control of fuzzy polynomial discrete-time nonlinear systems

CHAPTER OUTLINE

8.1 INTRODUCTION

The H_∞-control problem can be stated as finding a nonlinear controller such that 1) the internal state of the closed-loop system is stable and 2) the L_2 gain of the mapping from the exogenous input noise to the controlled output like tracking error or cost variables is minimized or guaranteed to be less or equal to a prescribed value. This problem has been solved by the dissipativity theory [1,2] and the nonlinear bounded real lemma [3,4] (see, e.g., [6,7], and [8]). Both of these approaches show that the existence of a solution to the nonlinear H_∞-control problem is in fact related to the solvability of the so-called Hamilton–Jacobi equation (HJE). Until now, there is no effective way of solving HJEs.

Over the past two decades, there has been rapidly growing interest in approximating a nonlinear system by the Takagi–Sugeno (TS) fuzzy model [9–34]. Based on this fuzzy model, a model-based fuzzy control was developed to stabilize the nonlinear system. This fuzzy modeling approach provides a powerful tool for modeling complex nonlinear systems. Unlike conventional modeling approaches where a single model is used to describe the global behavior of a systems, the TS modeling approach is essentially a multimodel approach in which simple submodels (typically, linear models) are combined to describe the global behavior of the system. Recently,

Analysis and Synthesis of Polynomial Discrete-Time Systems. DOI: 10.1016/B978-0-08-101901-6.00008-6

a polynomial fuzzy modeling has been proposed, which is a generalization of the TS fuzzy model and is more effective in representing nonlinear control systems [35].

The existing methods cannot guarantee a Lyapunov function to be a radially unbounded polynomial function, and hence the global stability cannot be assured. In this chapter, we propose a method guaranteeing to render a Lyapunov function that is a radially unbounded polynomial function. To the best of our knowledge, there exist no results yet on guaranteeing polynomial Lyapunov functions. With a polynomial Lyapunov function, the global stability of discrete-time polynomial fuzzy systems with a state feedback controller can be guaranteed. The problem of nonconvexity is solved by incorporating an integrator into the controller structure. In this study, this is developed for polynomial fuzzy systems in a more general framework to ensure global stability. Finally, sufficient conditions for the existence of a state H_∞ feedback controller are given in terms of polynomial matrix inequalities, which are solved via SOSTOOLS [36].

8.2 SYSTEM DESCRIPTION AND PRELIMINARIES

The discrete-time polynomial fuzzy system is described by the following rules:

Plant Rule i: IF $\upsilon_1(k)$ is J_1^i AND $\cdots$ AND $\upsilon_p(k)$ is J_p^i, THEN

$$\begin{aligned} x(k+1) &= \textstyle\sum_{i=1}^{r} \mu_i(\upsilon(k))\big\{A_i(x(k))\check{x}(k) + \Delta A_i(x(k))\check{x}(k) \\ &\quad + B_i(x(k))u(k) + \Delta B_i(x(k))u(k) + B_{w_i}(x(k))w(k)\big\} \qquad (8.1) \\ z(k) &= \textstyle\sum_{i=1}^{r} \mu_i(\upsilon(k))\big\{C_i(x(k))\check{x}(x(k)) + D_i(x(k))u(k)\big\}, \end{aligned}$$

where i denotes the ith fuzzy inference rule ($i = 1, \ldots, r$), $\upsilon_1(k), \ldots, \upsilon_p(k)$ are the premise variables, p is the number of premise variables, and $J_1^i, \ldots, J_p^i$ are the fuzzy terms. Furthermore, $x(k) \in \Re^n$, $u(k) \in \Re^m$, $w(k) \in \Re^l$, and $z(k) \in \Re^p$ are the state vector, control signals, disturbance, and vector of output signals related to the performance of the control system, respectively; $\check{x}(x(k)) \in \Re^N$ is a column vector whose entries are the monomials (in $x(k)$) such that $\check{x}(x(k)) = T(x(k)).x(k)$ where $T(x(k))$ is a polynomial transformation matrix from $x(k)$ to $\check{x}(x(k))$. $\check{x}(x(k))$ helps us to be more flexible in constructing the polynomial fuzzy model (8.1). The matrices $A_i(x(k))$, $B_i(x(k))$, $B_{w_i}(x(k))$, $C_i(x(k))$, and $D_i(x(k))$ have appropriate dimensions. Meanwhile, $\Delta A_i(x(k))$ and $\Delta B_i(x(k))$ represent the uncertainties of the form

$$\begin{aligned} &\begin{bmatrix} \Delta A_i(x(k)) & \Delta B_i(x(k)) \end{bmatrix} \\ &= H(x(k))F(x(k))\begin{bmatrix} E_{1i}(x(k)) & E_{2i}(x(k)) \end{bmatrix}, \end{aligned} \qquad (8.2)$$

where $H(x(k))$, $E_{1i}(x(k))$, and $E_{2i}(x(k))$ are known polynomial matrices of appropriate dimensions, and $F(x(k))$ satisfies

$$\left\| F^T(x(k))F(x(k)) \right\| < I. \tag{8.3}$$

By using a center-average defuzzifier, product inference, and singleton fuzzifier, we obtain the following global nonlinear model:

$$\begin{aligned} x_{k+1} &= A(\mu, x_k)\check{x}_k(x_k) + \Delta A(\mu, x_k)\check{x}_k(x_k) + B(\mu, x_k)u_k \\ &\quad + \Delta B(\mu, x_k)u_k + B_w(\mu, x_k)w_k, \\ z_k &= C(\mu, x_k)\check{x}_k(x_k) + D(\mu, x_k)u_k, \end{aligned} \tag{8.4}$$

where $\upsilon(k) = [\upsilon_1(k), \ldots, \upsilon_p(k)]$, $\chi_i(\upsilon(k)) = \prod_{t=1}^{p} J_t^i(\upsilon_t(k))$,

$$\mu_i(\upsilon(k)) = \frac{\chi_i(\upsilon(k))}{\sum_{\ell=1}^{r} \chi_\ell(\upsilon(k))} \in [0, 1], \ \sum_{i=1}^{r} \mu_i(\upsilon(k)) = 1,$$

$$A(\mu, x_k) = \sum_{i=1}^{r} \mu_i(\upsilon(k))A_i(x(k)), \ B(\mu, x_k) = \sum_{i=1}^{r} \mu_i(\upsilon(k))B_i(x(k)),$$

$$B_w(\mu, x_k) = \sum_{i=1}^{r} \mu_i(\upsilon(k))B_{w_i}(x(k)), \ C(\mu, x_k) = \sum_{i=1}^{r} \mu_i(\upsilon(k))C_i(x(k)),$$

$$\Delta A(\mu, x_k) = \sum_{i=1}^{r} \mu_i(\upsilon(k))\Delta A_i(x(k)),$$

$$\Delta B(\mu, x_k) = \sum_{i=1}^{r} \mu_i(\upsilon(k))\Delta B_i(x(k)),$$

$$E_1(\mu, x_k) = \sum_{i=1}^{r} \mu_i(\upsilon(k))E_{1i}(x(k)), \ E_2(\mu, x_k) = \sum_{i=1}^{r} \mu_i(\upsilon(k))E_{2i}(x(k)),$$

and $D(\mu, x_k) = \sum_{i=1}^{r} \mu_i(\upsilon(k))D_i(x(k))$.

To stabilize this system, the following polynomial fuzzy controller is employed:

$$\begin{aligned} x_c(k+1) &= x_c(k) + A_c(\mu, \check{x}_k, x_c), \\ u(k) &= x_c(k), \end{aligned} \tag{8.5}$$

where x_c is an integrator or controller state, and

$$A_c(\mu, \check{x}_k, x_c) = \sum_{i=1}^{r} \mu_i(\upsilon(k))A_{c_i}(\check{x}_k, x_c) \tag{8.6}$$

is the integrator input. To design the controller means to design $A_{c_i}(x_k, x_c)$ in such a way that the closed-loop system is stable.

Robust H_∞ Problem Formulation: Given a prescribed H_∞ performance $\gamma > 0$, design a nonlinear feedback controller (8.5) such that

$$\|z_k\|_{[0,\infty]} < \gamma^2 \|w_k\|_{[0,\infty]} \tag{8.7}$$

and system (8.4) with (8.5) is globally asymptotically stable for all admissible uncertainties.

Motivated by [37], we define the scaled system

$$\begin{aligned} \tilde{x}_{k+1} &= A(\mu, \tilde{x}_k)\check{x}_k + \left[\begin{array}{cc} B_w(\mu, \tilde{x}_k) & \frac{1}{\delta\gamma}\bar{H}(\tilde{x}_k) \end{array}\right]\tilde{w}_k + B(\mu, \tilde{x}_k)u_k, \\ z_k &= \left[\begin{array}{c} C(\mu, \tilde{x}_k) \\ \delta E_1(\mu, \tilde{x}_k) \end{array}\right]\check{x}_k(\tilde{x}_k) + \left[\begin{array}{c} D_u(\mu, \tilde{x}_k) \\ \delta E_2(\mu, \tilde{x}_k) \end{array}\right]u_k, \end{aligned} \tag{8.8}$$

where $\delta > 0$, $x_k \in \Re^n$ is the state vector, $\bar{H}(\tilde{x}_k) = \left[\begin{array}{cc} H(\tilde{x}_k) & H(\tilde{x}_k) \end{array}\right]$, $\tilde{w}_k = \left[\begin{array}{c} w_k \\ \delta\eta(k) \end{array}\right]$, and $\delta\eta(k) = F(x(k))\left[\begin{array}{c} E_1(\mu, x_k)x_k \\ E_2(\mu, x_k)u_k \end{array}\right]$.

Theorem 21. *Consider system (8.4) and controller (8.5). Property (8.7) holds for (8.4) with (8.5) for all admissible uncertainties if and only if there exists a positive constant $\delta >$, such that (8.7) holds for system (8.8) with the same controller.*

Proof. This theorem can be proven in the same way as in [37]. □

Lemma 5. *If there exist a symmetric polynomial matrix $P(x) > 0$ and a polynomial matrix $G(x)$, then $G^T(x) + G(x) - P(x) \le G^T(x)P^{-1}G(x)$.*

Proof. Since $P(x) > 0$, we have $(P(x) - G(x))^T P^{-1}(x)(P(x) - G(x)) \ge 0$. This implies $G^T(x) + G(x) - P(x) \le G^T(x)P^{-1}(x)G(x)$. □

8.3 MAIN RESULTS

In this section, we examine the global stabilization of discrete-time polynomial fuzzy systems (8.8) with the state feedback H_∞ control (8.5). System (8.8) with (8.5) is

$$\begin{aligned} \bar{x}_{k+1} &= \left(\hat{A}(\mu, \bar{x}_k) + \hat{B}(\mu, \bar{x}_k)\hat{A}_c(\mu, \hat{x}_k)\right)\hat{x}_k + \hat{B}_w(\mu, \bar{x}_k)\tilde{w}_k, \\ \tilde{z}_k &= \hat{C}(\mu, \bar{x}_k)\hat{x}_k, \end{aligned} \tag{8.9}$$

where $\hat{x}(k) = \begin{bmatrix}\check{x}_k^T(k) & x_c^T(k)\end{bmatrix}^T$, $\bar{x}(k) = \begin{bmatrix}\tilde{x}_k^T(k) & x_c^T(k)\end{bmatrix}^T$, $\hat{x}(k) = \hat{T}(\tilde{x}_k)\bar{x}(k)$,

$$\hat{T}(\tilde{x}_k) = diag\{T(\tilde{x}_k), 1\}, \; \hat{C}(\mu, \bar{x}_k) = \begin{bmatrix}\tilde{C}(\mu, \tilde{x}_k) & \tilde{D}(\mu, \tilde{x}_k)\end{bmatrix},$$

$$\tilde{C}(\mu, \tilde{x}_k) = \begin{bmatrix} C(\mu, \tilde{x}_k) \\ \delta E_1(\mu, \tilde{x}_k) \end{bmatrix}, \; \tilde{D}(\mu, \tilde{x}_k) = \begin{bmatrix} D(\mu, \tilde{x}_k) \\ \delta E_2(\mu, \tilde{x}_k) \end{bmatrix},$$

$$\hat{A}_c(\mu, \hat{x}_k) = A_c(\mu, \hat{x}_k) = \sum_{i=1}^{r} \mu_i(\upsilon(k)) A_{c_i}(\hat{x}_k),$$

$$\hat{A}(\mu, \bar{x}_k) = \begin{bmatrix} A(\mu, \tilde{x}_k) & B(\mu, \tilde{x}_k) \\ 0 & 1 \end{bmatrix}, \; \hat{B}_w(\mu, \bar{x}_k) = \begin{bmatrix} \tilde{B}_w(\mu, \tilde{x}_k) \\ 0 \end{bmatrix},$$

$$\hat{B}_w(\mu, \tilde{x}_k) = \begin{bmatrix} B_w(\mu, \tilde{x}_k) & \frac{1}{\delta}\bar{H}(\tilde{x}_k) \end{bmatrix}, \text{ and } \hat{B}(\mu, \bar{x}_k) = \begin{bmatrix} 0 \\ 1 \end{bmatrix}.$$

Sufficient conditions for the existence of a nonlinear H_∞ controller are given in the following theorem.

Theorem 22. *Given $\gamma > 0$ and $\delta > 0$, the polynomial fuzzy system (8.8) with controller (8.5) is globally asymptotically stable with prescribed H_∞ performance γ if there exists a positive definite matrix $\hat{P}$, a polynomial matrix $\hat{L}_i(\hat{x}_k)$, and lower unitriangular polynomial matrix $\hat{G}(\tilde{x}_k)$ such that the following conditions hold:*

$$\begin{cases} \hat{\Lambda}_{ii} > 0, & 1 \le i \le r, \\ \frac{1}{r-1}\hat{\Lambda}_{ii} + \frac{1}{2}(\hat{\Lambda}_{ij} + \hat{\Lambda}_{ji}) > 0, & 1 \le i \ne j \le r, \end{cases} \tag{8.10}$$

where

$$\hat{\Lambda}_{ij} = \begin{bmatrix} \hat{P} & \bullet & \bullet & \bullet \\ 0 & \gamma^2 I & \bullet & \bullet \\ \hat{T}(\tilde{x}_{k+1})\hat{\Phi}_{ij}(\hat{x}_k) & \hat{B}_{w_i} & \hat{G}(\tilde{x}_{k+1}) + \hat{G}^T(\tilde{x}_{k+1}) - \hat{P} & \bullet \\ \hat{C}_i\hat{G}(\tilde{x}_k) & 0 & 0 & I \end{bmatrix}, \tag{8.11}$$

$$\hat{\Phi}_{ij}(\hat{x}_k) = \hat{A}_i(\bar{x}_k)\hat{G}(\tilde{x}_k) + \hat{B}_i(\bar{x}_k)\hat{L}_j(\hat{x}_k), \tag{8.12}$$

$$\hat{G}(\tilde{x}_k)) = \begin{bmatrix} 1 & 0 & \cdots & 0 \\ \hat{G}_{21}(\tilde{x}_k) & 1 & 0 & 0 \\ \vdots & & \ddots & 0 \\ \hat{G}_{N1}(\tilde{x}_k) & \hat{G}_{N2}(\tilde{x}_k) & \cdots & 1 \end{bmatrix}, \tag{8.13}$$

$\hat{G}_{N1}(\tilde{x}_k)), \ldots, \hat{G}_{N2}(\tilde{x}_k))$ *are polynomial scalars, and the controller gain parameters can be obtained from* $\hat{G}(\tilde{x}_k)$ *and* $\hat{L}_i(\tilde{x}_k)$ *as*

$$\hat{A}_{c_i}(\hat{x}_k) = \hat{L}_i(\hat{x}_k)\hat{G}^{-1}(\tilde{x}_k)) \tag{8.14}$$

for $i = 1, \ldots, r$.

Proof. Consider the Lyapunov function

$$\hat{V}(\hat{x}_k) = \hat{x}_k^T \hat{G}^{-T}(\tilde{x}_k)\hat{P}\hat{G}^{-1}(\tilde{x}_k)\hat{x}_k. \tag{8.15}$$

Taking the forward difference of $\hat{V}(\hat{x}_k)$ gives

$$\begin{aligned}\Delta\hat{V}(\hat{x}_k) &= \hat{V}(\hat{x}_{k+1}) - \hat{V}(\hat{x}_k)\\ &= *\hat{P}\hat{G}^{-1}(\tilde{x}_{k+1})\hat{x}_{k+1} - *\hat{P}\hat{G}^{-1}(\tilde{x}_k)\hat{x}_k,\end{aligned} \tag{8.16}$$

By (8.9) the latter becomes

$$\begin{aligned}\Delta\hat{V}(\hat{x}_k) = *\hat{P}\hat{G}^{-1}(x_{k+1})\hat{T}(\tilde{x}_{k+1})\Big[\Big(\hat{A}(\mu,\hat{x}_k) + \hat{B}(\mu,\hat{x}_k)\hat{A}_c(\mu,\hat{x}_k)\Big)\hat{x}_k\\ + \hat{B}_w(\mu,\hat{x}_k)w_k\Big] - *\hat{P}\hat{G}^{-1}(\tilde{x}_k)\hat{x}_k.\end{aligned}$$

Adding and subtracting $-z_k^T z_k + \gamma^2 w_k^T w_k$ to and from $\Delta\hat{V}(\hat{x}_k)$, we obtain

$$\begin{aligned}\Delta\hat{V}(\hat{x}_k) &= *\hat{P}\hat{G}^{-1}(\tilde{x}_{k+1})\hat{T}(\tilde{x}_{k+1})\Big[\Big(\hat{A}(\mu,\hat{x}_k) + \hat{B}(\mu,\hat{x}_k)\hat{A}_c(\mu,\hat{x}_k)\Big)\hat{x}_k\\ &+ \hat{B}_w(\mu,\hat{x}_k)w_k\Big] - *\hat{P}\hat{G}^{-1}(\tilde{x}_k)\hat{x}_k + z_k^T z_k - \gamma^2 w_k^T w_k - z_k^T z_k + \gamma^2 w_k^T w_k\\ &= *\hat{P}\hat{G}^{-1}(\tilde{x}_{k+1})\hat{T}(\tilde{x}_{k+1})\Big[\Big(\hat{A}(\mu,\hat{x}_k) + \hat{B}(\mu,\hat{x}_k)\hat{A}_c(\mu,\hat{x}_k)\Big)\hat{x}_k\\ &+ \hat{B}_w(\mu,\hat{x}_k)w_k\Big] - *\hat{P}\hat{G}^{-1}(\tilde{x}_k)\hat{x}_k + \Big[\hat{C}(\mu,\hat{x}_k)\hat{x}_k\Big]^T\Big[\hat{C}(\mu,\hat{x}_k)\hat{x}_k\Big]\\ &- \gamma^2 w_k^T w_k - z_k^T z_k + \gamma^2 w_k^T w_k = X^T\Omega X - z_k^T z_k + \gamma^2 w_k^T w_k,\end{aligned}$$

where $X = \begin{bmatrix} \hat{x}_k^T & w_k^T \end{bmatrix}^T$ and

$$\Omega = *\hat{P}\hat{G}^{-1}(\tilde{x}_{k+1})\hat{T}(\tilde{x}_{k+1})\Theta_1(\mu,\hat{x}_k) + \Theta_2^T(\mu,\hat{x}_k)\Theta_2(\mu,\hat{x}_k) - \Pi$$

with

$$\Theta_1(\mu,\hat{x}_k) = \begin{bmatrix} \hat{A}(\mu,\hat{x}_k) + \hat{B}(\mu,\hat{x}_k)\hat{A}_c(\mu,\hat{x}_k) & \hat{B}_w(\mu,\hat{x}_k) \end{bmatrix}$$

and

$$\Theta_2(\mu,\hat{x}_k) = \hat{C}(\mu,\hat{x}_k) \text{ and } \Pi = \begin{bmatrix} \hat{G}^{-T}(\tilde{x}_k)\hat{P}\hat{G}^{-1}(\tilde{x}_k) & 0\\ 0 & \gamma^2 I \end{bmatrix}.$$

To have $\Delta\hat{V}(\hat{x}_k) < -z_k^T z_k + \gamma^2 w_k^T w_k$, we need $\Omega < 0$. Using the Schur complement on $-\Omega > 0$ results in

$$\begin{bmatrix} \Pi & \bullet & \bullet \\ \hat{T}(\tilde{x}_{k+1})\Theta_1(\mu,\hat{x}_k) & \hat{G}^T(\tilde{x}_{k+1})\hat{P}^{-1}\hat{G}(\tilde{x}_{k+1}) & \bullet \\ \Theta_2(\mu,\hat{x}_k) & 0 & I \end{bmatrix} > 0, \tag{8.17}$$

$$\begin{bmatrix} \hat{G}^{-T}(\tilde{x}_k)\hat{P}\hat{G}^{-1}(\tilde{x}_k) & \bullet & \bullet & \bullet \\ 0 & \gamma^2 I & \bullet & \bullet \\ \hat{A}(\mu.\hat{x}_k) + \hat{B}(\mu.\hat{x}_k)\hat{A}_c(\mu.\hat{x}_k) & \hat{B}_w(\mu.\hat{x}_k) & \hat{G}^T(\tilde{x}_{k+1})\hat{P}^{-1}\hat{G}(\tilde{x}_{k+1}) & \bullet \\ \hat{C}(\mu.\hat{x}_k) & 0 & 0 & I \end{bmatrix} > 0. \tag{8.18}$$

Then, multiplying (8.18) on the right by $diag\left\{\hat{G}^T(\tilde{x}_k), I, I, I\right\}$ and on the left by $diag\left\{\hat{G}(\tilde{x}_k), I, I, I\right\}$ and defining $\hat{L}(\mu,\hat{x}_k) = \hat{A}_c(\mu,\hat{x}_k)\hat{G}(\tilde{x}_k)$, we have

$$\begin{bmatrix} \hat{P} & \bullet & \bullet & \bullet \\ 0 & \gamma^2 I & \bullet & \bullet \\ \hat{T}(x_{k+1})\hat{\Phi}(\mu,\hat{x}_k) & \hat{B}_w(\mu,\hat{x}_k) & \hat{G}^T(\tilde{x}_{k+1})\hat{P}^{-1}\hat{G}(\tilde{x}_{k+1}) & \bullet \\ \hat{C}(\mu,\hat{x}_k)\hat{G}(\tilde{x}_k) & 0 & 0 & I \end{bmatrix} > 0, \tag{8.19}$$

where $\hat{\Phi}(\mu,\hat{x}_k) = \hat{A}(\mu,\hat{x}_k)\hat{G}(\tilde{x}_k) + \hat{B}(\mu,\hat{x}_k)\hat{L}(\mu,\hat{x}_k)$. Using Lemma 5, inequality (8.19) is satisfied if

$$\Lambda = \begin{bmatrix} \hat{P} & \bullet & \bullet & \bullet \\ 0 & \gamma^2 I & \bullet & \bullet \\ \hat{T}(\tilde{x}_{k+1})\hat{\Phi}(\mu.\hat{x}_k) & \hat{B}_w(\mu.\hat{x}_k) & \hat{G}(\tilde{x}_{k+1}) + \hat{G}^T(\tilde{x}_{k+1}) - \hat{P} & \bullet \\ \hat{C}(\mu,\hat{x}_k)\hat{G}(\tilde{x}_k) & 0 & 0 & I \end{bmatrix} > 0, \tag{8.20}$$

where $\Lambda = \sum_{i=1}^{r}\sum_{j=1}^{r}\mu_i(\upsilon(k))\mu_j(\upsilon(k))\Lambda_{ij}$. Suppose the conditions given in (8.10) are true, which implies that (8.20) is true. Now we have

$$\Delta\hat{V}(\hat{x}_k) < -z_k^T z_k + \gamma^2 w_k^T w_k. \tag{8.21}$$

Taking the summation from 0 to ∞ yields

$$\hat{V}(\hat{x}_\infty) - \hat{V}(\hat{x}_0) < -\sum_{k=0}^{\infty} z_k^T z_k + \gamma^2 \sum_{k=0}^{\infty} w_k^T w_k. \tag{8.22}$$

Since $\hat{V}(\hat{x}_0) = 0$ and $\hat{V}(\hat{x}_\infty) \geq 0$, we obtain

$$\sum_{k=0}^{\infty} z_k^T z_k < \gamma^2 \sum_{k=0}^{\infty} w_k^T w_k. \tag{8.23}$$

Hence (8.7) holds, and therefore the H_∞ performance is met.

To prove the stability, we set the disturbance $w_k = 0$. Therefore, system (8.8) with controller (8.5) is described as

$$\bar{x}_{k+1} = \left(\hat{A}(\mu, \bar{x}_k) + \hat{B}(\mu, \bar{x}_k)\hat{A}_c(\mu, \hat{x}_k)\right)\hat{x}_k. \tag{8.24}$$

Consider the same Lyapunov function

$$\hat{V}(\hat{x}_k) = \hat{x}_k^T \hat{G}^{-T}(\tilde{x}_k)\hat{P}\hat{G}^{-1}(\tilde{x}_k)\hat{x}_k. \tag{8.25}$$

Taking the forward difference of $\hat{V}(\hat{x}_k)$ gives

$$\begin{aligned} \Delta\hat{V}(\hat{x}_k) &= \hat{V}(\hat{x}_{k+1}) - \hat{V}(\hat{x}_k) \\ &= *\hat{P}\hat{G}^{-1}(\tilde{x}_{k+1})\hat{x}_{k+1} - *\hat{P}\hat{G}^{-1}(\tilde{x}_k)\hat{x}_k. \end{aligned} \tag{8.26}$$

By (8.9) the latter becomes

$$\begin{aligned} \Delta\hat{V}(\hat{x}_k) = *\hat{P}\hat{G}^{-1}(\tilde{x}_{k+1})\hat{T}(\tilde{x}_{k+1})\left[\left(\hat{A}(\mu, \bar{x}_k) + \hat{B}(\mu, \bar{x}_k)\hat{A}_c(\mu, \hat{x}_k)\right)\hat{x}_k\right] \\ - *\,\hat{P}\hat{G}^{-1}(\tilde{x}_k)\hat{x}_k. \end{aligned} \tag{8.27}$$

Conditions (8.10) imply

$$\begin{bmatrix} \hat{P} & \bullet \\ \hat{T}(\tilde{x}_{k+1})\hat{\Phi}(\mu, \hat{x}_k) & \hat{G}(\tilde{x}_{k+1}) + \hat{G}^T(\tilde{x}_{k+1}) - \hat{P} \end{bmatrix} > 0,$$

and hence system (8.24) is globally asymptotically stable. □

Remark 39. The introduction of an integrator into the controller renders

$$\hat{G}(\tilde{x}_{k+1}) = \hat{G}\left(A(\mu, \tilde{x}_k) + B(\mu, \tilde{x}_k)x_c(k)\right)$$

and

$$\hat{T}(\tilde{x}_{k+1}) = \hat{T}(A(\mu, \tilde{x}_k) + B(\mu, \tilde{x}_k)x_c(k)),$$

where $x_c(k)$ is the controller state. With this incorporation, the terms in $\hat{G}(\tilde{x}_{k+1})$ and $\hat{T}(\tilde{x}_{k+1})$ are convex. In other words, introducing an integrator in the controller makes the control matrix to be of the form $\hat{B}(\hat{x}_k) = [0 \quad 1]^T$, containing n zero rows related to system states $(\tilde{x}_k)$.

Remark 40. $\hat{G}(\tilde{x}_k)$ is lower unitriangular polynomial matrix. If there is no feasible solution, then it can be assumed as an upper unitriangular polynomial matrix without loss of generality. In this chapter, $\hat{G}^{-1}(\tilde{x}_k) = adj\left(\hat{G}(\tilde{x}_k)\right)$ is lower (upper) unitriangular polynomial matrix. Hence, the Lyapunov function is polynomial and radially unbounded, and so the global stability has been investigated.

The conditions given in Theorem 22 can be converted into SOS conditions as in the following corollary which can be solved by an SOS solver.

Corollary 12. *Given $\gamma > 0$ and $\delta > 0$, the polynomial fuzzy system (8.8) with controller (8.5) is globally asymptotically stable with the prescribed H_∞ performance γ if there exists a positive definite matrix $\hat{P}$, a polynomial matrix $\hat{L}_i(\hat{x}_k)$, a lower unitriangular polynomial matrix $\hat{G}(\tilde{x}_k)$, and polynomial scalars $\epsilon_{1i}(\hat{x}_k) > 0$ and $\epsilon_{2ij}(\hat{x}_k) > 0$ such that the following conditions hold:*

$$\begin{cases} v_1^T\left(\hat{\Lambda}_{ii} - \epsilon_{1i}(\hat{x}_k)I\right)v_1 & \text{is an SOS,} \\ & 1 \le i \le r, \\ v_2^T\left(\frac{1}{r-1}\hat{\Lambda}_{ii} + \frac{1}{2}(\hat{\Lambda}_{ij} + \hat{\Lambda}_{ji}) - \epsilon_{2ij}(\hat{x}_k)I\right)v_2 & \text{is an SOS,} \\ & 1 \le i \ne j \le r, \end{cases} \tag{8.28}$$

where

$$\hat{\Lambda}_{ij}(\hat{x}_k) = \begin{bmatrix} \hat{P} & \bullet & \bullet & \bullet \\ 0 & \gamma^2 I & \bullet & \bullet \\ \hat{T}(\tilde{x}_{k+1})\hat{\Phi}_{ij}(\hat{x}_k) & \hat{B}_{w_i}(\hat{x}_k) & \hat{G}(\tilde{x}_k) + \hat{G}^T(\tilde{x}_k) - \hat{P} & \bullet \\ \hat{C}_i(\hat{x}_k)\hat{G}(\tilde{x}_k) & 0 & 0 & I \end{bmatrix}, \tag{8.29}$$

$$\hat{\Phi}_{ij}(\hat{x}_k) = \hat{A}_i(\tilde{x}_k)\hat{G}(\tilde{x}_k) + \hat{B}_i(\hat{x}_k)\hat{L}_j(\hat{x}_k), \tag{8.30}$$

$$\hat{G}((\tilde{x}_k)) = \begin{bmatrix} 1 & 0 & \cdots & 0 \\ \hat{G}_{21}(\tilde{x}_k) & 1 & 0 & 0 \\ \vdots & & \ddots & 0 \\ \hat{G}_{N1}(\tilde{x}_k) & \hat{G}_{N2}(\tilde{x}_k) & \cdots & 1 \end{bmatrix}, \tag{8.31}$$

$\hat{G}_{N1}(\tilde{x}_k), \ldots, \hat{G}_{N2}(\tilde{x}_k)$ are polynomial scalars, v_1 and v_2 are vectors independent of x_k, and the controller gain parameters are obtained from $\hat{G}(\tilde{x}_k)$

and $\hat{L}_i(\hat{x}_k)$ *as*

$$\hat{A}_{c_i}(\hat{x}_k) = \hat{L}_i(\hat{x}_k)\hat{G}^{-1}(\tilde{x}_k) \tag{8.32}$$

for $i = 1, \ldots, r$.

8.4 SIMULATION EXAMPLES

Example 1. Consider the following discrete-time nonlinear system:

$$\begin{cases} x_1(k+1) &= (-\sin(x_1(k) + 0.05\Delta_a(k))\,x_1(k) \\ &\quad + (0.01 + 0.05\Delta_a(k))\,x_2(k)x_1(k) + 0.01u(k) + 0.01w(k), \\ x_2(k+1) &= (0.1 + 0.01\delta(k))\,x_1(k) + (0.5 + 0.01\delta(k))\,x_2(k) \\ &\quad +0.01u(k), \\ z(k) &= 0.01x_1(k) + 0.01x_2(k), \end{cases} \tag{8.33}$$

where $|\Delta_a(k)| \leq 1$ is the uncertain parameter. Using the technique described in [38] and taking the premise variable $\upsilon(k) = x_1(k)$, (8.33) can be expressed as

$$\begin{aligned} &\begin{bmatrix} x_1(k+1) \\ x_2(k+1) \end{bmatrix} \\ &= \mu_1(\upsilon)\left\{\left(\begin{bmatrix} -1 & 0.01x_1 \\ 0.1 & 0.5 \end{bmatrix} + \begin{bmatrix} 0.05x_2 & 0 \\ 0 & 0.01 \end{bmatrix}\delta\right)\begin{bmatrix} x_1 \\ x_2 \end{bmatrix}\right. \\ &\quad \left. + \begin{bmatrix} 0.01 \\ 0.01 \end{bmatrix}u + \begin{bmatrix} 0.1 \\ 0 \end{bmatrix}w\right\} \\ &\quad + \mu_2(\upsilon)\left\{\left(\begin{bmatrix} 1 & 0.01x_1 \\ 0.1 & 0.5 \end{bmatrix} + \begin{bmatrix} 0.05x_2 & 0 \\ 0 & 0.01 \end{bmatrix}\delta\right)\begin{bmatrix} x_1 \\ x_2 \end{bmatrix}\right. \\ &\quad \left. + \begin{bmatrix} 0.01 \\ 0.01 \end{bmatrix}u + \begin{bmatrix} 0.1 \\ 0 \end{bmatrix}w\right\}, \end{aligned} \tag{8.34}$$

$$z(k) = \mu_1(\upsilon)\,[0.01\;\;0.01]\begin{bmatrix} x_1 \\ x_2 \end{bmatrix} + \mu_2(\upsilon)\,[0.01\;\;0.01]\begin{bmatrix} x_1 \\ x_2 \end{bmatrix},$$

where $\mu_1(\upsilon(k)) = \frac{1}{2}(1 + \sin(\upsilon(k))$, $\mu_2(\upsilon(k)) = 1 - \mu_1(\upsilon(k))$, and $w(k)$ is the disturbance.

Before applying Corollary 12, we append an integrator into (8.34) and select $\hat{T}(x_k) = I$. Then system (8.34) can be represented in the form of (8.9), where $\bar{x}^T(k) = \begin{bmatrix} x_1^T(k) & x_2^T(k) & x_c^T(k) \end{bmatrix}$,

$$\hat{A}_1(\bar{x}(k)) = \begin{bmatrix} -1 & 0.01x_1(k) & 0.01 \\ 0.1 & 0.5 & 0.01 \\ 0 & 0 & 1 \end{bmatrix}, \quad \hat{B}_1(\bar{x}(k)) = \begin{bmatrix} 0 \\ 0 \\ 1 \end{bmatrix},$$

$$\hat{A}_2(\bar{x}(k)) = \begin{bmatrix} 1 & 0.01x_1(k) & 0.01 \\ 0.1 & 0.5 & 0.01 \\ 0 & 0 & 1 \end{bmatrix}, \quad \hat{B}_2(\bar{x}(k)) = \begin{bmatrix} 0 \\ 0 \\ 1 \end{bmatrix},$$

$$\hat{B}_{w1}(\bar{x}(k)) = \hat{B}_{w2}(\bar{x}(k)) = \begin{bmatrix} 0.01 & \frac{0.05x_2(k)}{\delta} & 0 & \frac{0.05x_2(k)}{\delta} & 0 \\ 0.01 & 0 & \frac{0.01}{\delta} & 0 & \frac{0.01}{\delta} \end{bmatrix},$$

and

$$\hat{C}_1(\bar{x}(k)) = \hat{C}_2(\bar{x}(k)) = \begin{bmatrix} 0.01 & 0.01 & 0 \\ \delta & 0 & 0 \\ 0 & \delta & 0 \end{bmatrix}.$$

We select $\delta = 1$, $\gamma = 1$, $\epsilon_1 = 0.01$, and $\epsilon_2 = 0.01$. The structures of $\hat{G}(x_k)$, $\hat{L}_1(\bar{x}_k)$, and $\hat{L}_2(\bar{x}_k)$ are chosen as

$$\hat{G}(x_k) = \begin{bmatrix} 1 & 0 & 0 \\ \hat{G}_{21}(x_k) & 1 & 0 \\ \hat{G}_{31}(x_k) & \hat{G}_{32}(x_k) & 1 \end{bmatrix},$$

$$\hat{L}_1(\bar{x}_k) = \begin{bmatrix} \hat{L}_{11}(\bar{x}_k) & \hat{L}_{12}(\bar{x}_k) & \hat{L}_{13}(\bar{x}_k) \end{bmatrix},$$

$$\hat{L}_2(\bar{x}_k) = \begin{bmatrix} \hat{L}_{21}(\bar{x}_k) & \hat{L}_{22}(\bar{x}_k) & \hat{L}_{23}(\bar{x}_k) \end{bmatrix}.$$

The degree of each polynomial element in polynomial matrices $\hat{G}(x_k)$ and $\hat{L}_i(\bar{x}_k)$ is selected to be 8 and 10, respectively. Applying Corollary 12 with these parameters, a feasible solution is obtained. The values of the polynomial elements in $\hat{G}(x_k)$, $\hat{L}_1(\bar{x}_k)$, and $\hat{L}_2(\bar{x}_k)$ are omitted here due to their sizes. Fig 8.1 shows the contour curves for the proposed Lyapunov function in the three subplots: (a) $\hat{V}(x_1, 0, z) = \alpha$, (b) $\hat{V}(0, x_2, z) = \alpha$, and (c) $\hat{V}(x_1, x_2, 0) = \alpha$ for $\alpha = 0.1, 1, 5, 10, 20, 30$. Note that, in all the plots in Fig 8.1, the contour curves are closed. Thus, if the states tend to infinity in any direction, then $\hat{V}$ increases and tends to infinity. Hence, the Lyapunov function is radially unbounded, and the closed-loop system is globally stable.

The obtained H_∞ performance is 9.36×10^{-4}, as shown in Fig 8.2, which is the ratio of the controlled output $z(k)$ energy to the disturbance $w(k)$ energy. This implies that the L_2 gain from the disturbance to the regulated output is

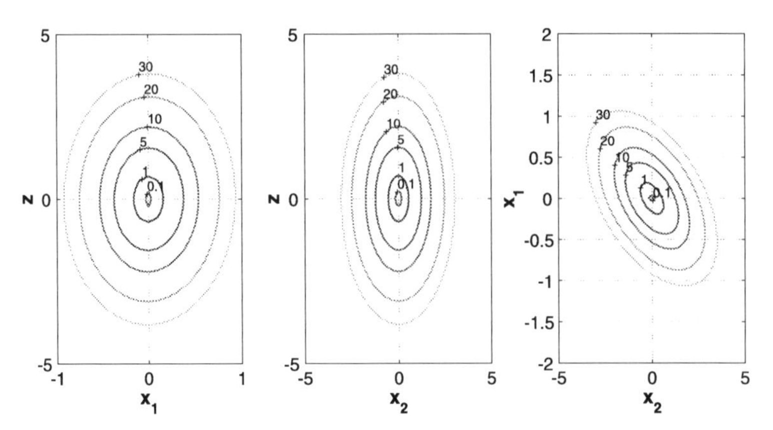

■ **FIGURE 8.1** Contours of the Lyapunov function. (A): $\hat{V}(x_1, 0, z)$, (B): $\hat{V}(0, x_2, z)$, and (C): $\hat{V}(x_1, x_2, 0)$.

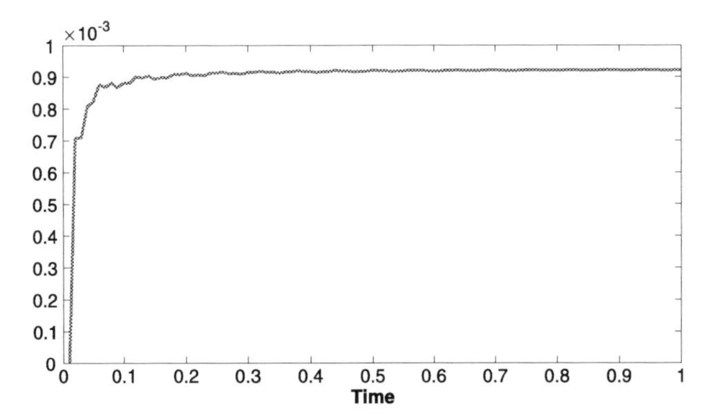

■ **FIGURE 8.2** Energy ratio ($\frac{\sum z^T z}{\sum \omega^T \omega}$).

no greater than $\gamma_{min} = 9.36 \times 10^{-4}$ and is much less than the prescribed value $\gamma = 1$.

Example 2. Consider the following discrete-time nonlinear pendulum system:

$$\begin{aligned} x_1(k+1) &= x_1(k) + 0.1x_2(k), \\ x_2(k+1) &= 0.1x_1(k)\sin(x_1(k) + 0.1x_2^2(k)x_1(k)(1 + \Delta_a(x_1(k), k)) \\ &\quad + x_2(k) + 0.1(1 + \Delta_b(x_1(k), k))u(k) + 0.1w(k), \\ z(k+1) &= x_1(k) + u(k), \end{aligned} \tag{8.35}$$

where $|\Delta_a(x_1(k), k)| \leq 1$ and $|\Delta_b(x_1(k), k)| \leq 1$ are uncertainties in the system. Employing [38] with the premise variable $\upsilon(k) = x_1(k)$, (8.35) can be expressed in the form (8.4) with

$$A_1(x(k)) = \begin{bmatrix} 1 & 0.1 \\ 0.1 & 0.1x_1(k)x_2(k) \end{bmatrix}, B_1(x(k)) = \begin{bmatrix} 0 \\ 0.1 \end{bmatrix},$$

$$B_{w1}(x(k)) = \begin{bmatrix} 0 \\ 0.1 \end{bmatrix},$$

$$A_2(x(k)) = \begin{bmatrix} 1 & 0.1 \\ -0.1 & 0.1x_1(k)x_2(k) \end{bmatrix}, B_2(x(k)) = \begin{bmatrix} 0 \\ 0.1 \end{bmatrix},$$

$$B_{w2}(x(k)) = \begin{bmatrix} 0 \\ 0.1 \end{bmatrix},$$

$$C_1(x(k)) = C_2(x(k)) = \begin{bmatrix} 1 & 0 \\ 0 & 0 \end{bmatrix}, D_1(x(k)) = D_2(x(k)) = \begin{bmatrix} 0 \\ 1 \end{bmatrix},$$

$$H(x(k)) = \begin{bmatrix} 0 & 0 \\ 0.1x_1(k)x_2(k) & 0.1 \end{bmatrix},$$

$$E_{11}(x(k)) = E_{12}(x(k)) = \begin{bmatrix} 0 & 1 \\ 0 & 0 \end{bmatrix},$$

$$F(x(k), k) = \begin{bmatrix} \Delta_a(x_1(k), k) & 0 \\ 0 & \Delta_b(x_1(k), k) \end{bmatrix},$$

$$E_{21}(x(k)) = E_{22}(x(k)) = \begin{bmatrix} 0 \\ 1 \end{bmatrix},$$

$\mu_1(\upsilon(k)) = \frac{1}{2}(1 + \sin(\upsilon(k)), \mu_2(\upsilon(k)) = 1 - \mu_1(\upsilon(k))$, and$-\frac{\pi}{2} \leq \upsilon(k) \leq \frac{\pi}{2}$.

Appending an integrator and applying Corollary 12 with $T(x_k) = I$, $\delta = 1$, $\gamma, \epsilon_1 = 0.01$, and $\epsilon_2 = 0.01$. The structure of $\hat{G}(x_k)$, $\hat{L}_1(\hat{x}_k)$, and $\hat{L}_2(\hat{x}_k)$ is given by

$$\hat{G}(x_k) = \begin{bmatrix} \hat{G}_{11} & 0 & 0 \\ \hat{G}_{21}(x_k) & \hat{G}_{22} & 0 \\ \hat{G}_{31}(x_k) & \hat{G}_{32}(x_k) & \hat{G}_{33} \end{bmatrix},$$

$$\hat{L}_1(\bar{x}_k) = \begin{bmatrix} \hat{L}_{11}(\bar{x}_k) & \hat{L}_{12}(\bar{x}_k) & \hat{L}_{13}(\bar{x}_k) \end{bmatrix},$$

$$\hat{L}_2(\bar{x}_k) = \begin{bmatrix} \hat{L}_{21}(\bar{x}_k) & \hat{L}_{22}(\bar{x}_k) & \hat{L}_{23}(\bar{x}_k) \end{bmatrix},$$

where $\bar{x}^T(k) = \begin{bmatrix} x_1^T(k) & x_2^T(k) & x_c^T(k) \end{bmatrix}$. The degree of each polynomial element in polynomial matrices $\hat{G}(x_k)$ and $\hat{L}_i(\bar{x}_k)$ is selected to be 12 and 15, respectively. With these parameters, a feasible solution is obtained. The values of the polynomial elements in $\hat{G}(x_k)$, $\hat{L}_1(\bar{x}_k)$, and $\hat{L}_2(\bar{x}_k)$ are

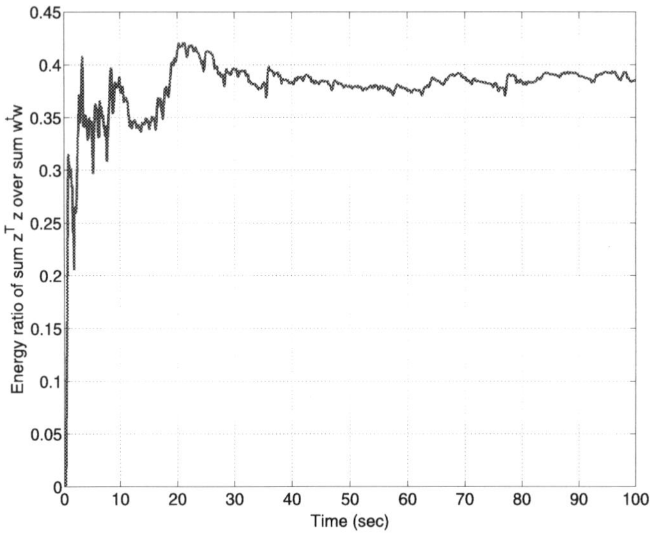

FIGURE 8.3 Energy ratio: $\frac{\sum z^T z}{\sum \omega^T \omega}$.

too large to be included here. The ratio of the estimation error energy to the noise energy is shown in Fig 8.3. It can be clearly seen from the figure that the energy ratio tends to be a constant value after 80 seconds, which is approximately 0.38. Hence, the γ value is equivalent to $\sqrt{0.38} \approx 0.6164$, which is less than the prescribed value $\gamma = 1$.

8.5 CONCLUSION

Based on the SOS approach, the robust H_∞ control of uncertain discrete-time polynomial fuzzy systems with H_∞ performance objective has been investigated. We have examined discrete-time polynomial fuzzy systems with norm-bounded uncertainties. The design conditions have been derived in terms of SOS constraints that can be numerically solved via the SOSTOOLS. The proposed control design procedures have been shown to yield a radially unbounded polynomial Lyapunov function, which ensures the global closed-loop stability.

REFERENCES

[1] T. Başar, P. Bernhard, H_∞-optimal control and related minimax design problems, in: Systems and Control: Foundations and Applications, Birkhäuser, Boston, 1991.

[2] J.A. Ball, J.W. Helton, H_∞ control for nonlinear plants: connection with differential games, in: Proc. 28th IEEE Conf. Decision Control, Tampa, FL, 1989, pp. 956–962.

[3] J.C. Willems, Dissipative dynamical systems. Part I: general theory, Archive for Rational Mechanics and Analysis 45 (1992) 321–351.

[4] D.J. Hill, P.J. Moylan, Dissipative dynamical systems: basic input–output and state properties, Journal of the Franklin Institute 309 (1980) 327–357.

[5] T. Basar, G.J. Olsder, Dynamic Noncooperative Game Theory, Academic Press, New York, 1982.

[6] A.J. van der Schaft, L_2-gain analysis of nonlinear systems and nonlinear state feedback H_∞ control, IEEE Transactions on Automatic Control 31 (1992) 770–784.

[7] A. Isidori, A. Astolfi, Disturbance attenuation and H_∞-control via measurement feedback in nonlinear systems, IEEE Transactions on Automatic Control 31 (1992) 1283–1293.

[8] A. Isidori, Feedback control of nonlinear systems, in: Proc. First European Control Conf., Grenoble, France, 1991, pp. 1001–1012.

[9] B.S. Chen, C.S. Tseng, H.J. Uang, Mixed H_∞ fuzzy output feedback control design for nonlinear dynamic systems: an LMI approach, IEEE Transactions on Fuzzy Systems 8 (Apr. 2000) 249–265.

[10] S.K. Nguang, P. Shi, Stabilization of a class of nonlinear time-delay systems using fuzzy models, in: Proc. Conf. Decision Control, Sydney, Australia, 2000, pp. 4415–4419.

[11] M. Teixeira, S.H. Zak, Stabilizing controller design for uncertain nonlinear systems using fuzzy models, IEEE Transactions on Fuzzy Systems 7 (Feb. 1999) 133–142.

[12] K. Tanaka, M. Sugeno, Stability analysis and design of fuzzy control systems, Fuzzy Sets and Systems 45 (1992) 135–156.

[13] K. Tanaka, T. Ikeda, H.O. Wang, Robust stabilization of a class of uncertain nonlinear systems via fuzzy control: quadratic stabilizability H control theory, and linear matrix inequality, IEEE Transactions on Fuzzy Systems 4 (Feb. 1996) 1–13.

[14] F. Rasool, D. Huang, S.K. Nguang, Robust H_∞ output feedback control of networked control systems with multiple quantizers, Journal of the Franklin Institute 349 (3) (2012) 1153–1173.

[15] S.K. Nguang, W. Assawinchaichote, P. Shi, Y. Shi, Robust H_∞ control design for uncertain fuzzy systems with Markovian jumps: an LMI approach, in: American Control Conference, 2005, pp. 1805–1810.

[16] S.K. Nguang, P. Zhang, S.X. Ding, Parity relation based fault estimation for nonlinear systems: an LMI approach, International Journal of Automation and Computing 4 (2) (2007) 164–168.

[17] S.K. Nguang, Comments on "Robust stabilization of uncertain input-delay systems by sliding mode control with delay compensation", Automatica 37 (10) (2001) 1677.

[18] S.K. Nguang, P. Shi, H_∞ output feedback control of fuzzy system models under sampled measurements, Computers and Mathematics With Applications 46 (5) (2003) 705–717.

[19] W. Assawinchaichote, S.K. Nguang, P. Shi, Fuzzy Control and Filter Design for Uncertain Fuzzy Systems, Springer, 2006.

[20] S.K. Nguang, P. Shi, On designing filters for uncertain sampled-data nonlinear systems, Systems & Control Letters 41 (5) (2000) 305–316.

[21] D. Huang, S.K. Nguang, Robust Control for Uncertain Networked Control Systems With Random Delays, Springer Science & Business Media, 2009.

[22] J. Zhang, A.K. Swain, S.K. Nguang, Robust sensor fault estimation scheme for satellite attitude control systems, Journal of the Franklin Institute 350 (9) (2013) 2581–2604.

[23] J. Zhang, P. Shi, J. Qiu, S.K. Nguang, A novel observer-based output feedback controller design for discrete-time fuzzy systems, IEEE Transactions on Fuzzy Systems 23 (1) (2015) 223–229.

[24] S.K. Nguang, P. Shi, Delay-dependent H_∞ filtering for uncertain time delay nonlinear systems: an LMI approach, IET Control Theory & Applications 1 (1) (2007) 133–140.

[25] Z. Hou, J. Luo, P. Shi, S.K. Nguang, Stochastic stability of Ito differential equations with semi-Markovian jump parameters, IEEE Transactions on Automatic Control 51 (8) (2006) 1383–1387.

[26] W. Assawinchaichote, S.K. Nguang, P. Shi, E.K. Boukas, H_∞ fuzzy state-feedback control design for nonlinear systems with stability constraints: an LMI approach, Mathematics and Computers in Simulation 78 (4) (2008) 514–531.

[27] S. Chae, S.K. Nguang, SOS based robust H_∞ fuzzy dynamic output feedback control of nonlinear networked control systems, IEEE Transactions on Cybernetics 44 (7) (2014) 1204–1213.

[28] F. Rasool, D. Huang, S.K. Nguang, Robust H_∞ output feedback control of discrete-time networked systems with limited information, Systems & Control Letters 60 (10) (2011) 845–853.

[29] S.K. Nguang, P. Shi, Stabilisation of a class of nonlinear time-delay systems using fuzzy models, in: Proceedings of the 39th IEEE Conference on Decision and Control, 2000, pp. 5–11.

[30] S. Saat, S.K. Nguang, Nonlinear H_∞ output feedback control with integrator for polynomial discrete-time systems, International Journal of Robust and Nonlinear Control 25 (2015) 1051–1065.

[31] Y. Zhang, P. Shi, S.K. Nguang, H.R. Karimi, Observer-based finite-time fuzzy H_∞ control for discrete-time systems with stochastic jumps and time-delays, Signal Processing 97 (2014) 252–261.

[32] S. Chae, F. Rasool, S.K. Nguang, A. Swain, Robust mode delay-dependent H_∞ control of discrete-time systems with random communication delays, IET Control Theory & Applications 4 (6) (2010) 936–944.

[33] Y. Zhang, P. Shi, S.K. Nguang, Y. Song, Robust finite-time H_∞ control for uncertain discrete-time singular systems with Markovian jumps, IET Control Theory & Applications 8 (12) (2014) 1105–1111.

[34] H.O. Wang, K. Tanaka, M.F. Griffin, An approach to fuzzy control of nonlinear systems: stability and design issues, IEEE Transactions on Fuzzy Systems 4 (Feb. 1996) 14–23.

[35] K. Tanaka, H. Yoshida, H. Ohtake, H.O. Wang, A sum of squares approach to modeling and control of nonlinear dynamical systems with polynomial fuzzy systems, IEEE Transactions on Fuzzy Systems 17 (4) (2009) 911–922.

[36] S. Prajna, A. Papachristodoulou, P. Parrilo, Introducing SOSTOOLS: a general purpose sum of squares programming solver, in: Proceedings of the 41st IEEE Conference on Decision and Control, 2002, vol. 1, Dec. 2002, pp. 741–746.

[37] S.K. Nguang, Robust nonlinear H_∞ output feedback control, IEEE Transactions on Automatic Control 41 (7) (1996) 1005–1007.

[38] K. Tanaka, H.O. Wang, Fuzzy Control Systems Design and Analysis: A Linear Matrix Inequality Approach, Wiley, 2001.

Chapter 9

Conclusion

CHAPTER OUTLINE

9.1 SUMMARY OF BOOK

This book proposes novel methodologies for controller synthesis and filter design for polynomial discrete-time systems. For the controller synthesis, state feedback controllers and output feedback controllers are designed with and without H_∞ performance. Polytopic uncertainties and norm-bounded uncertainties are considered in this book. To ensure that a convex solution to the control design problem for polynomial discrete-time systems could be rendered efficiently, an integrator is incorporated into controller structures. It is shown that by incorporating the integrator into the controller structures, original systems can be transformed into augmented systems, and the Lyapunov function can be selected such that its matrix is only dependent upon the original state. In doing so, a nonconvex controller design problem of polynomial discrete-time systems can be converted into a convex design problem in a less conservative way than available approaches. This consequently allows the problem to be solved via SDP. The integrator method is also applied to filter design for polynomial discrete-time systems. The effectiveness and advantages of the proposed design methodologies are verified by numerical examples in every chapter. The simulation results show that the proposed design methodologies can fulfill the prescribed performance requirement.

Some conservatism aspect of the proposed method is also explained accordingly in most of the chapters. Generally, the main problem of this integrator method is computational complexity. This problem is common to all methods that are based on augmented systems. However, in this research work, the problem becomes more severe because all the involved matrices are defined in polynomial forms. This consequently creates a large size of SD,

Analysis and Synthesis of Polynomial Discrete-Time Systems. DOI: 10.1016/B978-0-08-101901-6.00009-8

which requires a lot of memory space to solve the problem. This is the reason of selecting small sizes of systems as in our numerical examples so that the feasibility problem of the controller synthesis and filter design can be performed efficiently.

To clarify the approach used in this research work, seven technical chapters are provided. In Chapter 2, the problem of designing a controller for polynomial discrete-time systems is raised. Then, a novel method, called an integrator method is proposed to solve the problem in a less conservative way than the available approaches. Based on this integrator method, a nonlinear feedback control is tackled, and it is shown that the results can be extended to the robust control problem with uncertainties. The robust nonlinear H_∞ state feedback control problem for polynomial discrete-time systems is discussed in Chapter 3. Meanwhile, Chapter 4 presents the results of the robust filter problem for systems with the existence of norm-bounded uncertainties. The robust H_∞ filtering problem is considered in Chapter 5. A robust nonlinear output feedback controller is developed in Chapter 6, in which the stability and the H_∞ performance objectives must be satisfied. The previous chapters' control design methods for discrete-time fuzzy polynomial systems cannot guarantee their Lyapunov function to be a radially unbounded polynomial function, and hence the global stability cannot be assured. Therefore, in Chapter 7, we propose a control design method that guarantees a radially unbounded polynomial Lyapunov functions, which ensures the global stability. This result is extended to robust H_∞ framework in Chapter 8.

Here is a summary of the contributions of this book:

- The controller synthesis for polynomial discrete-time systems is considered. The controller designs are performed with and without the performance objective, i.e., H_∞ control.
- A less conservative design procedure of controller synthesis is obtained by incorporating an integrator into the controller structure.
- A possible method for solving the filtering problem for polynomial discrete-time systems is also given. This is delivered with the help of an integrator approach.
- The polytopic and norm-bounded uncertainties are considered. The methodologies for solving uncertain polynomial discrete-time systems with polytopic and norm-bounded uncertainties are provided.
- A global control design method that guarantees a radially unbounded polynomial Lyapunov functions ensuring the global stability is provided.

As a result, this book provides a less conservative design methodology for the controller synthesis and filter design of polynomial discrete-time systems and represents a valuable and meaningful contribution to the development of an SOS-SDP-based solution in the framework of polynomial discrete-time systems.

9.2 FUTURE RESEARCH WORK

In general, control of polynomial discrete-time systems still remains an open area, and lots of research work need to be conducted. Further research work, to name a few, could be directed to the following areas:

1. Time-delays system is one of the important problem in the framework of control systems engineering [1–9]. In fact, to the authors' knowledge, no result has been presented yet in the framework of controller synthesis for polynomial discrete-time systems with time delays. Therefore, it is interesting to consider the design of a controller for polynomial discrete-time systems with time delays. It is also desirable to see whether the incorporation of an integrator into the controller structure could help in solving the problem or not. The success of controller synthesis for such discrete time-delays systems could provide a potential methodology of the controller synthesis for the networked control systems in which the system or plant is represented by polynomial discrete-time systems.
2. It is noticeable from the book that the incorporation of an integrator into the controller structure leads to the computational burden [10]. This is due to the fact that a large number of sparse is created when solving the problem. This problem is quite common in the field of SOS programming, especially when using SOSTOOLS. Hence, research on reducing the number of sparse is the most interesting one to be done in the future. The reduction of this sparse will reduce the size of SDP and hence could provide a better future for this integrator method because it can also be applied to higher-order systems than the systems used in this research work.
3. In this book, the observer design is not performed. Therefore, for future research, it is highly recommended to study this observer design [11–15] for polynomial discrete-time systems. The other interesting consideration for future research work is involving the fault-tolerant problem [16] and [17] for polynomial discrete-time systems.
4. Another future research direction is Markovian jump polynomial discrete-time systems. Markovian jump systems, also known as hybrid systems with a state vector, consist of two components, the state (differential equation) and the mode (Markov process) [18–20].

REFERENCES

[1] F. Rasool, D. Huangm, S.K. Nguang, Robust H_∞ output feedback control of networked control systems with multiple quantizers, Journal of the Franklin Institute 349 (3) (2012) 1153–1173.

[2] S.K. Nguang, Comments on "Robust stabilization of uncertain input-delay systems by sliding mode control with delay compensation", Automatica 37 (10) (2001) 1677.

[3] Y. Zhang, P. Shi, S.K. Nguang, H.R. Karimi, Observer-based finite-time fuzzy H_∞ control for discrete-time systems with stochastic jumps and time-delays, Signal Processing 97 (2014) 252–261.

[4] S. Chae, F. Rasool, S.K. Nguang, A. Swain, Robust mode delay-dependent H_∞ control of discrete-time systems with random communication delays, IET Control Theory & Applications 4 (6) (2010) 936–944.

[5] S. Chae, S.K. Nguang, SOS based robust H_∞ fuzzy dynamic output feedback control of nonlinear networked control systems, IEEE Transactions on Cybernetics 44 (7) (2014) 1204–1213.

[6] F. Rasool, D. Huang, S.K. Nguang, Robust H_∞ output feedback control of discrete-time networked systems with limited information, Systems & Control Letters 60 (10) (2011) 845–853.

[7] S.K. Nguang, P. Shi, Stabilisation of a class of nonlinear time-delay systems using fuzzy models, in: Proceedings of the 39th IEEE Conference on Decision and Control, 2000, pp. 5–11.

[8] D. Huang, S.K. Nguang, Robust Control for Uncertain Networked Control Systems With Random Delays, Springer Science & Business Media, 2009.

[9] S.K. Nguang, P. Shi, Delay-dependent H_∞ filtering for uncertain time delay nonlinear systems: an LMI approach, IET Control Theory & Applications 1 (1) (2007) 133–140.

[10] S. Saat, S.K. Nguang, Nonlinear H_∞ output feedback control with integrator for polynomial discrete-time systems, International Journal of Robust and Nonlinear Control 25 (2015) 1051–1065.

[11] S.K. Nguang, W. Assawinchaichote, P. Shi, Y. Shi, Robust H_∞ control design for uncertain fuzzy systems with Markovian jumps: an LMI approach, in: American Control Conference, 2005, pp. 1805–1810.

[12] S.K. Nguang, P. Shi, H_∞ output feedback control of fuzzy system models under sampled measurements, Computers and Mathematics With Applications 46 (5) (2003) 705–717.

[13] S.K. Nguang, P. Shi, On designing filters for uncertain sampled-data nonlinear systems, Systems & Control Letters 41 (5) (2000) 305–316.

[14] W. Assawinchaichote, S.K. Nguang, P. Shi, Fuzzy Control and Filter Design for Uncertain Fuzzy Systems, Springer, 2006.

[15] J. Zhang, P. Shi, J. Qiu, S.K. Nguang, A novel observer-based output feedback controller design for discrete-time fuzzy systems, IEEE Transactions on Fuzzy Systems 23 (1) (2015) 223–229.

[16] S.K. Nguang, P. Zhang, S.X. Ding, Parity relation based fault estimation for nonlinear systems: an LMI approach, International Journal of Automation and Computing 4 (2) (2007) 164–168.

[17] J. Zhang, A.K. Swain, S.K. Nguang, Robust sensor fault estimation scheme for satellite attitude control systems, Journal of the Franklin Institute 350 (9) (2013) 2581–2604.

[18] Z. Hou, J. Luo, P. Shi, S.K. Nguang, Stochastic stability of Ito differential equations with semi-Markovian jump parameters, IEEE Transactions on Automatic Control 51 (8) (2006) 1383–1387.

[19] W. Assawinchaichote, S.K. Nguang, P. Shi, E.K. Boukas, H_∞ fuzzy state-feedback control design for nonlinear systems with stability constraints: an LMI approach, Mathematics and Computers in Simulation 78 (4) (2008) 514–531.

[20] Y. Zhang, P. Shi, S.K. Nguang, Y. Song, Robust finite-time H_∞ control for uncertain discrete-time singular systems with Markovian jumps, IET Control Theory & Applications 8 (12) (2014) 1105–1111.

Appendix A

Mathematical

In this section, we introduce some mathematical background knowledge applied throughout this research.

A.1 LINEAR MATRIX INEQUALITY (LMI)

Since the SOS inequality is in fact complementary from the LMI, it is necessary to understand the concept of LMI. In this section, we present a brief overview regarding the LMI theory.

Since the early 1990s, with the development of interior-point method methods for solving LMI problems, the LMI method has gained increased interest and emerged as a useful tool for solving a number of control problems: synthesis of gain-scheduled (parameter-varying) controllers, mixed-norm and multiobjective control design, hybrid dynamical systems, and fuzzy control. Three important factors that make LMI techniques appealing:

- A variety of design specifications and constraints can be expressed as LMIs.
- Once formulated in terms of LMIs, a problem can be solved exactly by efficient convex optimization algorithms ("LMI solvers").
- Whereas most problems with multiple constraints or objectives lack analytical solutions in terms of matrix equations, they often remain tractable in the LMI framework. This makes LMI-based design a valuable alternative to classical "analytical" methods.

For system and control perspective, the importance of LMI optimization stems from the fact that a wide variety of system and control problems can be recast as LMI problems. Therefore recasting a control problem as an LMI problem is equivalent to finding a "solution" to the original problem.

An LMI has the form

$$F(x) = F_0 + \sum_{i=1}^{m} x_i F_i > 0, \tag{A.1}$$

where $x \in R^m$ is the variable to be determined, and symmetric matrices $F_i = F_i^T \in R^{n \times m}, i = 0, \ldots, m$, are given. The inequality symbol in (A.1) means that $F(x)$ is positive definite, i.e., $u^T F(x)u > 0$ for all nonzero $u \in R^n$.

Even though this canonical expression (A.1) is generic, LMIs rarely arise in this form in control applications. Instead, structured representation of LMIs is often used. For instance, the expression $A^T P + PA < 0$ in the Lyapunov inequality is explicitly described as a function of the matrix variable P, and A is the given matrix. In addition to saving notation, the structured representation may lead to more efficient computation.

A.2 THE SCHUR COMPLEMENT

The Schur complement is standard in the LMI context. The basic idea is as follows: The LMI

$$\begin{bmatrix} Q(x) & S(x) \\ S^T(x) & R(x) \end{bmatrix} > 0, \qquad \text{(A.2)}$$

where $Q(x) = Q^T(x)$, $R(x) = R^T(x)$, and $S(x)$ depend affinely on x, is equivalent to

$$R(x) > 0, \qquad Q(x) - S(x)R^{-1}(x)S^T(x) > 0. \qquad \text{(A.3)}$$

In other words, the system of nonlinear inequalities (A.3) can be represented as the LMI (A.2).

Index

Printed in the United States
By Bookmasters